T0195263

BIRDS & FLOWERS

Also by Jeff Ollerton:

Plant–Pollinator Interactions: From Specialization to Generalization (with Nickolas M. Waser [eds]; 2006)

Pollinators & Pollination: Nature and Society (Pelagic, 2021)

BIRDS & FLOWERS

An Intimate 50 Million Year Relationship

JEFF OLLERTON

PELAGIC PUBLISHING

First published in 2024 by
Pelagic Publishing
20–22 Wenlock Road
London N1 7GU, UK

www.pelagicpublishing.com

Birds & Flowers: An Intimate 50 Million Year Relationship

https://doi.org/10.53061/FFQO3130

British Library Cataloguing in Publication Data
A catalogue record for this book is available from the British Library

ISBN 978-1-78427-451-1 Hardback
ISBN 978-1-78427-452-8 ePub
ISBN 978-1-78427-453-5 PDF

Cover photo: Cape Weaver & *Strelitzia reginae* by Stephen Valentine,
www.instagram.com/fromstephensstudio/

Typeset in Minion Pro by S4Carlisle Publishing Services, Chennai, India

Printed and bound in Great Britain by TJ Books Limited, Padstow, Cornwall

'While it is but rarely that an ornithologist also possesses sufficient competence in botany to be able to conduct research of this nature without the aid of a specialist, and vice versa, an intimate co-operation between the two is clearly indicated for obtaining optimum results.'

Sálim Moizuddin Abdul Ali (1932)

This book is dedicated to the world's birders and botanisers: may your binoculars never fog up, may you never lose your hand lens, and may the natural world always be there to inspire your conversations.

Contents

	Introduction: Encounters with birds and flowers	1
1.	Origins of a partnership *Understanding 50 million years of bird and flower evolution*	7
2.	Surprising variety *The astounding diversity of pollinating birds*	18
3.	Keeping it in the family *Accounts of the different groups of bird pollinators*	27
4.	A flower's point of view *How many plants are bird-pollinated, and where are they found?*	43
5.	In the eye of the beholder *What do bird flowers look like?*	57
6.	Goods and services *The enticements given to birds for pollinating flowers*	66
7.	Misaligned interests *The ongoing conflicts between flowers and birds*	80
8.	Senses and sensitivities *How bird brains shape the flowers that they pollinate*	92
9.	Codependent connections *Networks of interacting flowers and birds*	106
10.	Hitchhikers, drunks and killers *The other actors in the network and how they affect the main players*	116
11.	The limits to specialisation *How 'specialised' are the relationships between birds and flowers?*	126

12. Islands in the sea, islands in the sky 141
Isolation, in oceans or in mountains, results in some remarkable interactions

13. The curious case of Europe 155
Why did we believe that Europe had no bird-pollinated flowers?

14. 'After the Manner of Bees' 167
The origins of our understanding of birds as pollinators, and their cultural associations

15. Feathers and fruits 185
Birds as pollinators of edible wild plants and domesticated crops

16. Urban flowers for urban birds 197
Bird pollination in cities and gardens

17. Bad birds and feral flowers 208
The impact of invasive species

18. What escapes the eye 220
The decline and extinction of bird–flower relationships

19. The restoration of hope 233
People as conservationists of birds and their flowers

Species names 247

Sources and further reading 253

Acknowledgements 291

Index 293

INTRODUCTION

Encounters with birds and flowers

The wind whipped through the branches of the tall, weeping bottlebrush, alternately revealing and obscuring the sunbirds that were clamouring around the drooping scarlet inflorescences. Sunbird identification can be difficult even under the best of circumstances, depending on the light and how it reflects from the iridescent feathers. Today, it was harder than ever. Was that a green cap I spotted? Perhaps with a red throat? Maybe an Amethyst Sunbird? What of the other birds that were clearly different? Two days later the wind had dropped and we were able to confirm at least four species: Amethyst Sunbird (our hunch was correct), plus Scarlet-chested, Variable and Tacazze Sunbirds. All attracted to the nectar-abundant flowers of a tree that evolved thousands of kilometres away and which would normally sustain populations of Australian honeyeaters.

Those observations were made in the garden of a hotel on the slopes of Mount Kenya, and if they prove anything it's that flower-visiting birds often don't much care what they forage on for nectar. Judging by the abundant dry seed capsules on the bottlebrush, the tree was also not too fussy.

Birds and flowers are, I would suggest, the two aspects of the natural world with which people are most familiar. Both can be extremely colourful and both, with some exceptions, usually confine their appearances and activities to the daylight hours. For this reason, 'birders' and 'botanisers' tend to be (along with entomologists) the most numerous of naturalists, the wildlife enthusiasts one is most likely to encounter out in the field or online in the various forums and social media. It seems surprising, then, that no one has written a book that provides a global overview of how birds and flowers interact. There have, of course, been plenty of books about hummingbirds, and at least two devoted to sunbirds. There have also been numerous volumes dedicated to pollinators and pollination, my own book in 2021 being just the latest addition to a literature stretching back more than 200 years. My purpose in writing *Birds & Flowers*, then, is to bring together the global diversity of these two distinct, but often interdependent, groups. Am I speaking about the organisms or the people? Actually both!

I have a long and deep interest in the natural world, an interest that stretches back to my earliest childhood memories, and which has shaped and directed my adult professional and personal life. But let me be honest from the beginning. I'm not an ornithologist, an avian expert who can tell you the name of every bird at 100 paces based purely on the sounds they make. Nor am I a botanist who can give you a trait-by-trait run-down of all of the features of leaves, stems and flowers that fit a given plant into its genus or family. In both cases I'll make a pretty good stab at identification, but that's not where my specialism lies. What I am is an ecologist who has, for over 30 years, been fascinated by the ways in which different types of organisms interact with one another and how those relationships between species have evolved, and continue to evolve, over time. These associations carry huge importance in the world's ecosystems, as ecologically vital players and as the drivers and objects of evolutionary novelty. In addition, as a source of cultural inspiration, with a history stretching back millennia, they are part of the ongoing relationship between humanity and the rest of nature, as seen in our ideas, religious and secular symbols, art and literature. And of course our activities on this planet threaten both the natural and the cultural heritage, and so cannot be ignored. Writing this book is therefore, in a sense, an attempt to reclaim birds as pollinators, to rescue them from being mere novelties of (especially Neotropical) natural history, and to explore the nature and significance of all their relationships with flowers.

During my career I have been lucky enough to travel widely, often carrying out field work or observing natural history in various parts of North and South America and Africa, in China and Nepal, in Australia, New Zealand and the Canary Islands, and of course in the UK and Denmark and other areas of Europe, from Scandinavia to the Mediterranean. That's an extraordinary privilege for a scientist and naturalist who seeks to understand the similarities and differences to be found across the different regions of the biosphere. Each of the chapters in this book includes examples of my own encounters with birds and flowers that draw on these experiences, as well as work published by others who are actively studying these topics.

The book is structured according to three themes, although they are to some extent interwoven. We begin with the evolution and subsequent diversification of the relationships between flowers and birds. We then move on to consider what we know (and, importantly, what we do not know) about the details and intricacies of the ecology of bird–flower interactions. Finally, we bring in a third dimension, examining the relationships between birds, flowers and humans.

There are numerous ways in which birds interact with flowers: for example, many species, including various pigeons and finches, exploit them for food; male bowerbirds use them as ornaments to decorate their bowers; berrypeckers even seem to have a medicinal use for them. Although I deal with some of these in later chapters, the main focus of this book is on how birds ensure flowering plant reproduction by acting as pollinators.

In Chapter 1 I set the scene for how bird pollination of flowers came about. Pollination is simply the ecological process of transferring pollen, containing male sex cells, from one flower to arrive at the stigma, which is the receptive female surface, of another flower. In essence, it's no different from sex between animals such as birds, except that usually a third party is involved to facilitate the transfer. In around 90% of the flowering plants, that third party is typically an insect such as a bee, fly or wasp, or a vertebrate like a bat, lizard or bird. It's a relationship that seems very familiar but which constantly surprises us when we start to dig into the details.

Chapters 2 and 3 assess the amazing abundance and diversity of bird pollinators, which goes far beyond the hummingbirds, sunbirds and honeyeaters with which you might be familiar. Parrots? Of course. Pigeons and doves? Why yes. Warblers and woodpeckers? I think that you will be pleasantly surprised!

In Chapters 4 and 5 I look at this diversity from the flower's point of view, considering the types of flowers that are bird-pollinated, which are surprisingly varied in their colours, shapes, sizes and other traits, and don't always conform to our expectations. Chapter 6 considers the nature of the interaction between birds and flowers, the ways in which birds are rewarded for their services, and the how flowers engage in the art of pollen placement on the various parts of a bird's body, including feathers and beaks, tongues and feet. In Chapter 7 we discuss how the particular interests of the birds and their flowers are misaligned – it's certainly not cooperation between the species, the animals and the plants having each evolved to exploit the world around them in their own interests.

Chapter 8 considers the sensory capabilities of birds, especially their remarkable vision and capacity to learn, how these have been exploited by flowers, and how this in turn has influenced floral evolution. There's also a discussion of something that all species (and comedians) know: timing is everything. For flowers this means knowing when to be active, whilst for birds it's a matter of timing daily and seasonal rhythms to the whims of their botanical partners.

Chapter 9 deals with the interactions between flowers and birds as networks of co-occurring and sometimes co-evolving species, while Chapter 10 – rather dramatically entitled 'Hitchhikers, drunks and killers' – brings in some of the other organisms that are mixed up in these relationships, from yeasts in nectar to predatory spiders and mantids. I bring this theme to a close in Chapter 11 by looking at the nature of specialised interactions between birds and flowers and considering an important but often neglected question – what does 'specialised' even mean in this context?

In Chapters 12 and 13 I narrow the geographical scope and consider first of all the importance of both oceanic islands and sky islands (isolated mountains) for understanding how bird and flower interactions evolve, with a special focus on the archipelagos of the Galápagos and Canary Islands. The latter links us directly to Europe and the question of why that continent is such an anomaly when it comes to bird-pollinated flowers and flower-pollinating birds.

The final few chapters look at birds and flowers from a human point of view, beginning with a historical and cultural perspective in Chapter 14. It may seem strange to look back in time to think about how we first acknowledged and then studied the surprising (to Europeans at least) fact that birds act as pollinators, and what they mean culturally and symbolically to indigenous peoples and society more widely. But sometimes we have to look backwards and sideways to understand the future. The relatively few, but regionally important, examples of bird pollination in agriculture are explored in Chapter 15, which leads us into the importance of our towns, cities, and gardens for birds and their flowers (or flowers and their birds, if you prefer) in Chapter 16. Chapter 17 then considers what happens when some of these 'bad birds and feral flowers' become invasive. Sometimes they have negative effects on ecosystems beyond their natural range, but also they may evolve in order to adapt to their new habitats.

The final two chapters, 18 and 19, reflect on how interactions between birds and flowers have changed during the Anthropocene, the species and interactions that have been lost, but also how people, organisations and communities are working hard to conserve ecosystems and restore hope in what sometimes seem to be desperate times for biodiversity. Human influences on the planet have greatly reduced the diversity and abundance of pollinating birds and their flowers, and climate change may fundamentally alter current relationships between these organisms. But committed people all over the globe are working hard to preserve and restore nature, and it's not too late to turn this around.

The informal bird names used in this book follow the Cornell Lab of Ornithology's *Birds of the World* online resource (birdsoftheworld.org/bow). I've also used informal names for plants and the other animals that feature in the book, where such names exist. As far as possible, I have not used scientific names in the text, but these can all be found in the species list at the end of the book. In the interests of readability, I have also avoided littering the text with formal citations of the many books, articles and websites that I've consulted. Instead, each chapter is supported by a list of sources and further reading.

The friendship and collegiality of my fellow scientists has been a big part of my life, and one of the reasons why my work has been so rewarding and so much fun. Many of these influential colleagues you'll meet in the chapters that follow, others I have thanked in the acknowledgements section at the end of the book. But it really all began with my PhD supervisor Andrew Lack, whose gift of a book sparked my initial fascination. In 1993 I dedicated my PhD thesis to my family with the call to travel: 'Australia, here we come!' And then, a few weeks after the thesis was completed, we set off on a journey that would take us first to Hong Kong, then on to the Great Southern Land, and back to the UK via New Zealand and California. To receive a field guide to the birds of Australia from Andrew was welcome preparatory reading for my first trip to the tropics. Although the focus of my research was more on flowers and their insect pollinators, I took the hint and tried to pay close attention to the birds as well. I never anticipated that the gift would eventually culminate in this book. I have a lot to thank Andrew for.

Bird pollination is an active and expanding area of research and there is a great deal of information and ideas that I could not fit into the book. But if this inspires you to read further and explore the topic in more detail, I'll consider my mission accomplished. It's an immensely varied and complex body of written work that I'm drawing from, in some cases going back several centuries. I hope I've done that corpus justice – and of course, any errors of fact are mine to be owned.

But to truly start our story of these ecological partners we need to go far back in time to consider what birds and flowers really are and how they originated. So settle in, we're heading to the Jurassic.

CHAPTER 1

Origins of a partnership

In the lush forests of the Jurassic period, perhaps 170 million years ago, a group of small-bodied dinosaurs began to evolve a set of novel traits that allowed them to exploit a semi-arboreal lifestyle that included the ability to glide between trees. A key feature of these dinosaurs was a layer of complex structures that covered most of their bodies, making this gliding more efficient, and at the same time providing thermal insulation and a biological canvas on which to display colours and patterns to attract mates, startle predators, or sneakily camouflage their bodies. The evolution of wings, clothed with these feathers, culminated in the 11,000 or so species that we now recognise as the only living dinosaurs: the birds.

Although aerial flight has been lost in several unrelated groups, including penguins and the ratites (ostriches and their relatives), the ability to fly was an important development that allowed the birds to raise their young in places that were physically off limits to most predators, such as sheer cliffs and the thin tips of tree branches. Not only that, they also gained access to sources of food that were previously out of reach, such as high-flying insects.

The insects had themselves evolved flight around 250 million years earlier, during the Devonian period, and this had likewise allowed them to diversify into new ecological niches that exploited the organisms with which they shared their primeval world. Amongst these other groups were the seed plants, with which the insects had begun an ecological partnership that traded sugar-rich fluids for sexual favours.

The earliest insect pollinators were initially interacting with various groups of non-flowering seed plants, collectively termed gymnosperms. However, at the end of the Jurassic, some 145 million years ago (or possibly much earlier, the debate still rages) a distinct group of seed plants appeared that had an unmatched ability to evolve into 'endless forms most beautiful and most wonderful', to borrow Charles Darwin's phrase. These flowering plants, or angiosperms, continued to diversify during the Cretaceous and remained diverse even after the mass extinction event about 66 million years ago that took out all of the dinosaurs except the birds, as well as many other groups of plants and animals. Angiosperms have therefore long been, and remain, the most diverse and dominant plants on the planet. Flowering plants in turn had formed partnerships with a diverse range of insects that we would recognise today as the principal groups of pollinators, including bees, flies, butterflies and moths, and beetles. And of course birds, who came late to the party, as we'll see, but have hung on as flamboyant and dedicated revellers.

And herein lies a profound mystery: why should flowers evolve interactions with pollinating birds at all? The world has never been short of insects, though conservation scientists warn us that there are far fewer now owing to the impacts humans have on the natural world. Insects are reliable flower visitors whose life histories are often intimately tied to the flowers which provide them with sugary nectar laced with minerals, amino acids and complex organic compounds. They also got there first, establishing mutually exploitative relationships with flowers tens of millions of years before birds entered the scene. One possibility is that there's a hidden chapter within the story of pollinators and pollination. In that scenario, small non-avian dinosaurs, or perhaps the smallest pterosaur species, were feeding on nectar and pollen of some of the early, larger-flowered angiosperms before birds. This set the scene for birds to move into that feeding niche following the end-Cretaceous mass extinction, which resulted in the demise of those other groups.

This is highly speculative and there's not (yet) any proof that this occurred. Indeed, writing in 1987, palaeontologist James Farlow was of the opinion that it was 'unlikely that many (any?) species of herbivorous dinosaurs fed on pollen or nectar'. If I've learned one thing in my career it's that, when it comes to the natural world, never say 'never'. James was writing at a time when the existence of lizard pollination was hardly known about, and yet now we recognise it as important for some plants and for the lizards that feed on nectar. Although lizards are not dinosaurs, nonetheless we should be cautious about jumping to definite conclusions about the lifestyles of ancient and extinct species.

But back to the original question: if insects are so effective as pollinators, and were undoubtedly there at the very beginning of flower evolution, why should these flowers shift to birds to aid their reproduction? In fact, there are some disadvantages to relying on birds. They have superior cognitive skills to most insects, and if it's possible to cheat in the relationship, by taking nectar without picking up or depositing pollen, they surely will find a way. That's especially true if carrying pollen has a cost to the birds, for example by interfering with how feathers function so that the bird must spend more time grooming, or if it is faster to access nectar by bypassing the sexual parts of the flower, thus saving energy. In addition, the rather large size of most birds compared to insects means that many flowers have, in turn, to evolve to be larger in order to accommodate birds' bodies, or at least their beaks. Large flowers, in turn, cost a plant energy and resources to build and maintain, reducing what's available for growth and seed production. There are, however, advantages to having birds move your pollen around, as we'll see later.

Although it's tempting to think of associations such as the pollination of flowers by insects and birds as being a 'partnership' and describe them as 'mutualistic', in fact these interactions are, like all ecological relationships, essentially exploitative. If a flower produces nectar, pollen or some other reward, and the flower visitor in turn moves pollen between flowers, then the species are mutually exploiting one another. It's often the case, however, that the flowers offer no reward or that the animal visitors fail to pick up or deposit pollen. Rewardlessness and reward thievery are the understandable outcomes of ecological systems in which exploitation of one species by another is a fundamental biological rule: any individual in a population that can cheat, thus avoiding the costs of providing a reward or service, will be at an advantage. We'll explore this further in Chapter 7 when we consider the misaligned interests of birds and flowers.

Depending on the species involved and the outcome of the interaction for both players, ecologists give these interactions names such as predation, herbivory, parasitism, mutualism, commensalism and so forth. But the boundaries between these different types of interaction are a bit blurred, and in fact they constitute a seamless tapestry in which it's not always possible to distinguish between, say, mutualism and parasitism. Birds are no different to any other group of animals in this regard, and we can find examples of all of these ecological interactions in the life histories of members of the Aves, the taxonomic class to which birds belong. Although the fossil evidence is limited, and in many cases we must make informed guesses based on largely skeletal remains, we can be certain that the earliest modern birds have exploited other species, and been exploited by them, ever since they evolved. Some of those fossils are extremely informative about the early exploitation of flowers by birds.

Birds and flowers in the fossil record

Palaeontologists can infer a lot from the shapes of bones and beaks. In 2013, for example, Polish scientists describing the fossil remains of *Resoviaornis jamrozi* concluded that the slim shape of its beak meant that it was probably an insect-eating (insectivorous) or fruit-eating (frugivorous) bird, whilst its legs, which were relatively long, meant that it was largely a ground dweller. In contrast, the extinct North American genus *Teratornis* included two species that were larger than modern condors with huge beaks that hint at these birds being aggressive predators of smaller mammals and carrion feeders on the carcasses of megafauna such as mammoths, although at least one species may have been predominantly a fish eater.

What about flower feeding? Is there evidence of that interaction to be found in the layers of rock that document ancient life on Earth? The answer is yes, but these fossils are incredibly uncommon. Fossils of birds that belong to the major families of avian pollinators are more common, so let's consider those first.

Honeyeater fossils are known from across the current range of these birds, with specimens described from Australia, New Zealand and Pacific islands including Hawaii. Most of these date to the Quaternary period and so are no more that about 2.5 million years old and often considerably younger. However, in 2016 Walter Boles from the Australian Museum in Sydney described a set of long lower leg bones from geological deposits in northwestern Queensland. To avian anatomy experts these are highly distinctive in the honeyeaters, and so Walter was able to attribute the bones to the family, if not to genus. Several species are involved, ranging in age from the Pliocene (5.3 to 2.6 million years ago) to the Middle to Late Miocene (about 16 to 5 million years ago).

In other words, even the oldest of these fossil honeyeaters are relatively recent in terms of bird evolution. However, we can push the timescale of birds as flower visitors back much further when we consider another major family of avian pollinators.

Hummingbirds are such a distinct group and they have a high reliance on flowers, such that any fossil hummer is likely to be indirect evidence of flower visitation. Fossil hummingbirds have been found, but the oldest of them are not where you'd expect, in the Americas, but in Europe. In 2004 German palaeontologist Gerald Mayr described such a bird from specimens collected in a clay pit in the south of the country. The fossil was dated to the Early Oligocene epoch, around 30 million years ago, and Gerald gave it the apt name of *Eurotrochilus inexpectatus* – 'the unexpected European hummingbird'. Subsequently, a second species of *Eurotrochilus* was discovered in Poland, and specimens from France may represent a third. In their time these were widespread, relatively abundant birds, and all the evidence suggests that they were flower feeders. The birds' beaks, although no more than 20 millimetres in length, were elongated relative to the size of the skull, and close examination of these small fossils indicates that they had all the necessary skeletal apparatus for sustained hovering. They were also tiny, in fact similar to the Bee Hummingbird from Cuba, which is the smallest known bird, only 5–6 centimetres in length and often weighing less than 2 grams.

These are by far the oldest specimens found of hummingbird ancestors, but they are still not the oldest fossils that provide us with evidence of flower feeding. That is attributable, once again, to Gerald Mayr, and his colleague Volker Wilde, and their description of a Middle Eocene fossil from the Messel Shale in Germany, one of the most important fossil assemblages in Europe. Messel fossils date to around 48 million years ago and document, in stunningly preserved detail, the plants and animals that ended up in a small lake surrounded by tropical forest, over a period of some 800,000 years. The fossil in question, named *Pumiliornis tessellatus*, was about the size of a wren but with a more elongated beak. Its relationship to living groups of birds is unclear because in some aspects of its anatomy it is similar to shorebirds such as the plovers, lapwings and avocets, whilst in other respects it recalls the crakes and rails. To make matters more complex, the feet of *Pumiliornis* are more like those of pigeons and doves. This combination of characters inspired the species name *tessellatus*, from the Latin word for 'mosaic'.

Regardless of its evolutionary affinities, a specimen of the bird (only the third to be found) was described in which the stomach contents were clearly preserved. Within the gut were some insect remains that could not be identified, plus, more excitingly, a rather large amount of pollen. All of this led Gerald and Volker to conclude that the bird was a flower visitor that probably fed mainly on nectar plus occasional insects (and perhaps fruit) in the tropical forests that existed in what was to become Europe, tens of millions of years ago. In the view of the authors, the substantial quantity of pollen was unlikely to have been ingested with the insects (for example pollen-collecting bees, which are also known from the Messel Shale), given that so few insect remains were found. The authors go further and speculate:

> As *Pumiliornis* does not belong to any of the modern groups of flower-visiting birds, the origin of [bird pollination] in some angiosperm

> lineages may have predated that of their extant avian pollinators.

They may well be correct, but there is another possible interpretation: the bird could have plucked and consumed whole flowers from a plant, a feeding strategy that ecologists refer to as florivory. The petals and associated tissues would have been easily digested, leaving behind the pollen grains with their tough external walls. Until we find further specimens with associated external pollen, we may never know if this was a pollinator. What's certain, however, is that *Pumiliornis* represents the earliest known interaction between a bird and flowers.

So it is clear that relationships between birds and flowers go back many millions of years. But why did those relationships evolve? And why did it reach the point where birds are important pollinators for many species of plant? What advantages does it convey to the birds? There are several possible explanations.

Why did birds start to visit flowers?

Describing birds as 'predators' or 'herbivores' or 'scavengers' or 'insectivores' or 'nectar feeders' does not do justice to avian diversity and the complexity of feeding behaviours. Most birds, in fact, shift between feeding on different types of food. However, it is at the nexus of two modes of feeding – insectivory and herbivory – that we may find the origin of behaviours which gave rise to the exploitation of flowers by birds, and of birds by flowers. In fact, the kinds of interactions that were probably the precursor to bird pollination of flowers can be seen in living species, in our own gardens.

In the house where my wife Karin and I used to live in Northampton, UK, a large greengage tree (a type of plum) each year provided us with an abundance of sweet, yellowish fruit that we ate straight from the tree or turned into jams and fruit crumbles. That's not all the tree provided, however. In the winter we watched the birds that used the bare branches as a perch, whilst in the early spring the white, fragrant flowers illuminated an otherwise dull corner of the garden. These flowers attracted not only the bees and flies that were in turn responsible for the fruit later in the season, but also birds. Common Wood-Pigeons massacred flowers on the highest and most inaccessible branches, tearing them off and swallowing them whole. Such florivory is not uncommon in birds, and we'll encounter it again later in the book. Another group of birds, the tits, were also attracted to the flowers but for a different reason: to feed on the small insects that accumulate in and around them. Watching Eurasian Blue Tits delicately probing greengage flowers to access their prey was always a highlight of spring in our garden. Further afield I would see them on other early flowering species within the rose family, such as Cherry Plum, and on willows alongside leaf warblers such as Common Chiffchaffs. The faces of these birds were often dusted in pollen, though their role as pollinators was probably minor compared to the bees and flies that were far more frequent and reliable flower visitors for these trees (but as I note in Chapter 13, the experiments required to test this have yet to be done).

Which of these two scenarios, browsing on flowers or predation of flower-visiting insects, is the more likely precursor to bird pollination? Florivory often leads to the destruction of most or all of the floral tissue. Any flower treated in such a way clearly cannot reproduce, and for that reason I think this is an unlikely path to specialised bird pollination, though there may be a role for this behaviour in the reproduction of some mass-flowering trees. But with insectivores visiting flowers to prey upon insects we start to see the kind of probing behaviour that could, more than 50 million years ago, have evolved into bird pollination. Key to this is not only the insect feeding of such small birds, but also the flexible feeding strategies of such species, which sometimes will incorporate nectar into their diets. As we will later see, there are flowers that have evolved to exploit tits, warblers and other non-specialised songbirds as their main pollinators.

There is also a clue to be found when we consider the evolutionary relationships between specialist nectar-feeding birds and their closest, non-flower-feeding relatives. Hummingbirds for example are most closely related to the swifts and the treeswifts, all species of which are obligate insect feeders. Sunbirds are passerines, and their closest relatives are the flowerpeckers, which despite their name, often feed on fruit. It's notable that precisely none of the 'specialist' nectar-feeding birds subsists purely on a diet of nectar. Even the most extreme specialised species, for example the Sword-billed Hummingbird that I'll discuss in Chapter 11, includes insects in its diet.

My own view is that nectar feeding (nectarivory) and therefore pollination can evolve via multiple pathways – frugivory and insectivory, as well as herbivory – depending on the group of birds under consideration. The subtitle of this book promises to tell the story of a 50-million-year-old relationship. As we saw above, the oldest fossils that have so far been discovered are around 48 million years old, so we have a couple of million years of artistic licence. However, my suspicion is that future fossil finds will push back the origin of bird pollination even further – because there are potentially more than 100 million years, from the earliest origins of birds, in which such relationships could have evolved. The birds that act as pollinators of flowers today are extremely diverse, and flower visitation has evolved as a behaviour many times in the past. Who knows what some of the extinct birds got up to? If they displayed anything like the diversity, flexibility and sophistication evident in modern bird–flower relationships, there is an enormous amount that we almost certainly will never know about these past interactions. The best the fossils can do is to hint at what has gone before.

Partnerships between flowers and pollinators are vital to plant reproduction and therefore to the way in which habitats such as grasslands and forests have sustained their integrity for tens of millions of years. Without pollinators to ensure seed production, most terrestrial ecosystems would fail.

And partnerships of a different kind are also important. For the scientists who study plant–pollinator interactions, collaboration with like-minded researchers from around the world has a vital role to play.

Going to California

The hummingbird interrupted our conversation. Darting from flower to flower, the tiny bird drained each blossom of nectar before moving on to the next, and the next, and the next. It flew and sounded like no other bird I had ever encountered. Of course, I had seen hummingbirds in natural history documentaries, who hasn't? But to watch one in real life, to see its precise hovering at close quarters and hear the insect-like *brrrzzz* of its wings, was to realise just how special these birds are. Not only that, but the flowers it was feeding on were the crimson blossoms of a planted eucalyptus tree. I had travelled halfway around the world, spending six months in Australia and one in New Zealand, before arriving in the USA, and it felt as though several elements of my journey had come together as a native North American bird foraged on the flowers of an Australian tree. All too soon the bird was gone, burning through the nectar sugars to fuel its flight as it sped off to find yet more flowers on which to feed.

Our conversation continued. I was sitting in the shade of the eucalyptus enjoying an *al fresco* lunch bathed in California sunshine, happy to be talking science with two other ecologists who were to play an important role in my professional development: Nick Waser and Mary Price. Coincidentally, they had also recently returned from Australia. In fact, all of us had been based at the same institution, Macquarie University in Sydney. But when I was on campus, Nick and Mary were in the field studying desert rodents. And when I was in the field, trying to hunt down the elusive pollinators of some rainforest climbers, they were back in the laboratory. So it was not until we met up at their home institution, the University of California at Riverside, that we'd finally had a chance to talk. We compared notes about our experiences in Australia, about the observations that we had made of pollinators and their interactions with flowers, and how generalist plants (those visited by a wide diversity of different pollinators) were under-appreciated by pollination ecologists. Many people in our area of research tended to focus on specialised flowers as their object of study, because they are more tractable to experimentation and tell a straightforward story of flower evolution in the face of a dedicated pollinator.

This was 1994, and two years later a paper that had its origins in my conversation with Nick and Mary was published in the journal *Ecology* as 'Generalization in pollination systems, and why it matters'. As I recalled in my book *Pollinators & Pollination*, that paper made something of a splash in the literature and divided opinion amongst our peers. But as I read back through it now, more than 25 years later, I realise that even then we were challenging some of the received wisdom about the role of birds as pollinators of flowers, and the ways in which flowers evolve relationships with the avian partners that service their reproductive needs.

Since that meeting beneath the eucalyptus, my partnerships with collaborators such as Nick, Mary, and many others have led me to study hummingbirds in Guyana, Peru and Brazil and sunbirds in Tanzania, Kenya and South Africa. I've encountered honeyeaters and lorikeets in Australia and a diverse set of non-typical pollinating birds, including warblers, tits and finches, in the Canary Islands and in Nepal. All of these birds, and others that you'll encounter in the next chapter, visit flowers to feed on nectar and sometimes pollen, and in the process instigate the reproduction of a plant.

CHAPTER 2

Surprising variety

Holding fast to a vertical cliff face, the tall, colourful inflorescences of the Canary Island Foxglove reached out into space, silhouetted against the cloudy sky. Viewing the plants through binoculars, I could just about make out the individual orange-red flowers, each one a perfect combination of form and colour, and each one containing a large nectar reward. Above the wind that seems to constantly blow through Tenerife's Anaga Mountains I could hear the insistent *cheerp* of birds communicating with one another and see the occasional bustle of small dark bodies in the dense, surrounding vegetation. Abruptly, one bird emerged and flew over to a foxglove, landing on the inflorescence, gripping the vertical stem tightly. As it sank its head into a succession of flowers, my hands held the binoculars tightly and I started to count … one … two … three … four … flower visits. At each visit the bird probed deeply, and I could just about make out the conjunction of the bird's head with the sexual parts of the flower. Turning, I grinned at my companions, ecstatic with my first view of a most unlikely bird pollinator: the Canary Island Chiffchaff.

Probably no other group of birds is more associated with visiting flowers than hummingbirds of the kind that concluded the previous chapter. They are the spangled troubadours of nectar feeding *par eminence*, intimately associated with blooms and the nectar that they provide, and pervasive in culture through art, literature and music, as we'll see in Chapter 14. And yet the 352 described species of hummingbirds represent just a fraction of the total number of birds that regularly or occasionally visit flowers. There are some which are only distantly related to the hummers, but almost equally nectar-dependent, such as sunbirds and honeyeaters, plus there's a smorgasbord of 'generalist' birds that are occasional, but sometimes important, pollinators, such as the Canary Island Chiffchaff. But how many different types of birds are we talking about? How many species? One of my tasks when researching this book was to find this out. I was helped along the way by one particular publication – a survey by Eugenie Regan and colleagues entitled 'Global trends in the status of bird and mammal pollinators'. For that study, the researchers had combed the scientific literature to produce a database of all the birds known to be pollinators, plus the bats, rodents, monkeys and other mammals, and then categorised them according to International Union for Conservation of Nature (IUCN) Red List criteria (something I will come back to in Chapter 18).

This was a great starting point for me, but since its publication in 2015 there had been other previously undocumented bird species recognised as pollinators. Also, Eugenie's criteria for including birds and mammals were very strict and there were whole families which I knew visited flowers and which were likely to be pollinators, but which were not included in the database.

So I scoured the published literature for examples of birds that visit flowers and are probably pollinators in at least some contexts. I also took note of photographs and videos that popped up on the internet and social media, which were sometimes very revealing. For example, the image on Wikipedia of the Indian Pied Starling (or Myna) visiting the flowers of a coral tree hints at what Indian birders already knew – it's a species that regularly visits flowers for nectar, even if that behaviour has never been formally studied or described by scientists.

Putting all of this together is an ongoing process. But at the time of writing I've found evidence that almost 1,390 bird species visit flowers in ways that suggest they might be pollinators – in that they do not (always) damage floral reproductive organs or consume the whole flower, or they have been observed with pollen on their bodies.

The generally accepted figure for the number of living or recently extinct birds is around 11,000 species. At the time of writing, the Cornell Lab of Ornithology's *Birds of the World* site lists

10,906 species, for example, while the IOC *World Bird List* (motto: 'Wisdom begins with putting the right name on a thing') names 11,001 living and 160 extinct species.

Based on those estimates, at least 12.5% of the world's bird species visit flowers. The real figure is probably higher, because there are many birds which we suspect may feed on nectar or pollen that have hardly been studied, especially in the wet tropics. Therefore I wouldn't be surprised if subsequent research shows that as many as 15% or 20% of bird species are flower visitors and have the potential to act as pollinators.

This is impressive enough, but when we look at higher taxonomic levels things become really fascinating. To appreciate what I discovered we need to take a little detour into the confusing world of how we put names to living things (taxonomy) and how we group these organisms together into related taxa (systematics). I know from decades of teaching biology students that taxonomy and systematics can be a turn-off for most people, who consider it a dry science filled with complicated and unpronounceable names. But it's a vital part of any story about ecology and evolution, so bear with me, I'll do my best to make this interesting!

In the last 20 years, new fossil finds and emerging molecular technologies have allowed us to reassess the science of bird systematics. For anyone wishing to go deeper into this I can recommend John Reilly's excellent book *The Ascent of Birds*. But bear in mind that there are conflicting views on many aspects of bird taxonomy and not all scientists agree even on fundamental questions such as the number of living bird species. A study in 2016, led by George Barrowclough from the Department of Ornithology at the American Museum of Natural History in New York, suggested that the true avian diversity may be over 20,000, double the generally accepted number. Not every ornithologist agrees with this, so there are major disagreements at this fundamental level, even before we ask questions about how these birds are related to one another. When it comes to classifying birds, or indeed any group of living organisms, taxonomic arguments are rife. I don't have a pigeon in this particular race or any other involving bird taxonomy, so if some of what I present below offends your systematic sensibilities, rest assured that I'm simply going with what seems to be the consensus rather than trying to feather my own particular intellectual nest. But that's enough bird puns for now. Let's dive into avian classification, and why it's important in the context of birds and flowers.

A brief primer in bird taxonomy

Taxonomy can be a challenging science for the uninitiated. In my experience, remembering the sequence of taxonomic categories, and the fact that each level in the sequence is contained within the previous higher level, like a set of nested boxes, is half the battle won. In my undergraduate lectures in biodiversity I use a mnemonic to help the class remember the different taxonomic levels:

King Prawn Curry Only Food That Gives Satisfaction

The initial letters (which are sometimes capitalised, but often not) remind my students that a species is a member of a genus, that these genera in turn are classified into tribes, which themselves belong in families. Those families can be grouped into orders, which are divisions of classes. Classes can be categorised into phyla, and each phylum belongs in a kingdom. I illustrate this by looking at some of the ingredients for this dish, such as prawns, rice and onions, showing how each is taxonomically classified within this system. The mnemonic seems to work, and years later graduates who have taken that class sometimes quote it back at me.

What about birds? Where do they fit? Well of course birds are animals, so they are members of the kingdom Animalia. They are also vertebrates, and therefore belong to the phylum Chordata. When it comes to class, birds are classified as Aves. This class contains a set of subgroups that broadly divide the birds along the lines of their long evolutionary history. The groups separate out the 'paleognaths' (ostriches, emus, kiwis and relatives) from the rest of the birds, which we term the 'neognaths'. Within the latter group, we see finer distinctions between the landfowl and waterfowl, and the rest of the birds. So it goes on, more finely dividing these groups and separating them on the basis of features of their anatomy and their DNA.

Eventually we get to the level of the order. There are reckoned to be about 44 living orders of birds, the names of which are all standardised to end with '–iformes', providing a helpful reminder of where we are located within the bird taxonomy. To give you an idea of how distinct these orders are, we're considering birds as different from one another as a Bald Eagle (Accipitriformes) is from a Herring Gull (Charadriiformes) or a Mute Swan (Anseriformes) is from a Bohemian Waxwing (Passeriformes). This last order is colloquially termed the 'perching birds' or 'passerines' and this one group constitutes 60% of living bird species.

Diving deeper still into the orders, we get to the rank of family. Family names of birds end in the suffix '–idae', again just to keep things standardised. If we explore this within just the passerines we find families as distinct as, say, birds of paradise (Paradisaeidae), wrens (Troglodytidae), wagtails and pipits (Motacillidae) and sparrows (Passeridae). Things get a little more complicated when you have distinct but closely related families; for example, 'owls' may be members of one of two families – the true owls (Strigidae) or the barn-owls (Tytonidae). There are also some surprises when we start to consider the relationships of these groups in terms of their DNA, not just their anatomy. For example, the Bald Eagle mentioned above is not at all closely related to any of the falcons, which belong in a completely different order (Falconiformes) that has more in common with the parrots and the passerines than it does with any of the other so-called 'birds of prey'.

There are various taxonomic levels below that of family, such as subfamily, tribe, and subtribe. But the next major one that concerns us is the rank of genus. Genus (or generic) names are the first of the two components that make up the scientific name of a species. Assuming that you're not an AI reading this, then you belong to the genus *Homo* and the species *sapiens* – *Homo sapiens*. Note that whatever you might read in a newspaper, this is the correct convention for presenting scientific names, with a capital initial for the genus, lowercase initial for the species, and written in *italics*.

A fairly large family such as the sunbirds and spiderhunters (Nectariniidae) contains multiple genera; in fact the 143 described species are divided into 16 genera, such as *Cyanomitra* and *Chalcomitra*, each with seven species, *Anthreptes*, which contains 14 species, and so on. Below this level we might also consider subspecies. So for example the Western Violet-backed Sunbird *Anthreptes longuemarei* has four subspecies in Africa, including *Anthreptes longuemarei longuemarei* in part of its northern range and *Anthreptes longuemarei angolensis* in areas further south. These subspecies names reflect the fact that different populations of the birds may have distinctive plumage, or calls, or behaviours, which in turn reflect differences at the genetic level that indicate they are evolving in different ways.

The latter point is crucial here, because in its essence taxonomy should tell us something about the evolutionary history (or 'phylogeny') of the species or genus or family concerned. It doesn't always, but that's due at least in part to the tensions between classical taxonomy (with its roots in the eighteenth century), which emphasises the physical appearance and behaviour of the bird, and the more modern approach of molecular phylogenetics, which looks deep within the DNA. But for the purposes of the story that we are developing, assessing the diversity of pollinating birds at a family level tells us something very important: an estimate of the number of times that birds have evolved this nectar-feeding habit. Here's the logic. The 363 species of hummingbirds (Trochilidae) share a common ancestor that flew in the forests of (possibly) Europe tens of millions of years ago. We can be fairly confident that nectar feeding evolved only once in that ancestral lineage, and then was retained by the members of the family as they diversified throughout the Americas. Likewise with the sunbirds (Nectariniidae) and the honeyeaters (Meliphagidae): each of these families represents a single evolution of nectar feeding. So that's at least three times that nectar feeding has evolved as a lifestyle for birds. Flower pollination as an outcome of that lifestyle has evolved much more frequently, as we'll see in Chapter 4.

Of course, assessing the number of times that nectar feeding has evolved is a little more complicated than this, and it depends on how closely related two families are. To give an example, although we classify them as different families, the sunbirds share a distant ancestor with the flowerpeckers (Dicaeidae), and species in both groups feed on nectar. So it is possible that nectarivory evolved just once in that ancestor and persisted into the birds that we now classify into two families. But these cases are rare: most of the families of flower-feeding birds that we are going to discuss are not closely related and so almost certainly evolved their potential as pollinators independently of one another. We'll meet these families in the next chapter, but before we do I want to clear up a persistent misconception that's often encountered when we read about bird pollination.

To hover or to perch?

Anyone who has watched a small insectivorous bird such as a Goldcrest gleaning for insects around the crown of a tree, or a kestrel scanning grassland vegetation for prey, hanging motionless in the sky, knows that the ability to hover is not unique to any one bird group. The hummingbirds, however, are the only family that have the necessary anatomy and sensory acuity to make this a mode of regular flight. It's energy-expensive, though, and if these birds can feed whilst perching, they will do so, as you can see even if you do not live in an area where hummingbirds naturally live. The live webcam of hummingbird feeders from the Sachatamia Lodge in Ecuador is a treat for anyone who loves these birds – and indeed nectar-feeding bats, which make an appearance at the feeders in the early hours of the Ecuadorian morning. Watch those birds and you will see that all of them perch to drink the sugar water from the feeders, hovering only to chase away rivals or to move to a different feeder.

It's curious, then, that many of the flowers on which hummingbirds naturally feed (that is to say, excluding those plants such as the Bird of Paradise that have been introduced from other parts of the world) do not offer their bird pollinators a place to perch. The flowers of hummingbird-adapted plants frequently dangle into space or emerge as thin-stemmed bouquets on which even the smallest hummers cannot gain a foothold, or are orientated such that hovering is the only option (see Plate 1). These plants force the birds to work for their rewards, perhaps because it is advantageous to the plant to keep the birds moving. In this respect, the flowers are exploiting the unique form of flight of hummingbirds. In contrast, it's always been thought that sunbird-adapted flowers provide some kind of firm perch on which the birds can stand while feeding (Plate 2). This perch may be the stem of the inflorescence or a specially adapted structure, such as the highly adapted perch discovered by Bruce Anderson and colleagues in some South African members of the iris family.

It's not always a clear-cut distinction, however. There are some hummingbird-pollinated flowers that require the birds to stand, especially at high elevations in the Andes (Plate 3). Likewise, sunbirds will sometimes hover whilst feeding. That was first studied in detail by Sjirk Geerts and Anton Pauw when they assessed how these birds took nectar from flowers of the invasive Tree Tobacco in South Africa, a species that is normally hummingbird-pollinated in its native range (Chapter 17). Non-specialist birds sometimes hover when feeding at flowers, too. Our study of birds taking nectar from rhododendron flowers in the Himalayas, which I'll discuss in more detail in Chapter 12, included observations of a leaf warbler hovering to feed, and this is clearly not a unique observation. In a 2013 review, Petra Wester documented examples of hovering while feeding on nectar in more than 80 non-hummingbird species belonging to 11 Old World families of birds. Most of those were sunbirds, and Petra concluded that 'the distinction between specialist and generalist nectarivorous birds is more adequate than that between hovering hummingbirds and perching passerines'. This distinction is something that I will return to in Chapter 11.

Recently it's been documented by Štěpán Janeček that there are indeed tropical African plants that require their sunbirds to hover rather than perch. As with the hummingbirds, it would appear to make evolutionary sense to force birds to move constantly between flowers by hovering rather than allowing them to take their time at flowers, thus moving pollen more effectively. However, I'm unaware of any studies that have shown that hummingbirds spend less time per flower than other groups of flower-visiting birds. Sunbirds, for instance, are fairly swift consumers of nectar. As always when we look beneath the surface of our expectations of nature, we find that things are more complex than first documented, and that we may not have the data to test some of our new assertions.

Whether they hover or perch, flower-feeding birds are diverse in shape, size and reliance on nectar. That diversity, as we shall see, reflects the order, family or genus of birds to which a species belongs. The number of living families of birds, and the names that we give to those families, like all aspects of avian taxonomy, are hotly debated. But currently the IOC *World Bird List* includes 44 orders, divided into 253 families, in turn split into 2,376 genera. How many of those families contain birds that act as pollinators? My estimate will, I think, surprise you. It certainly surprised me.

CHAPTER 3

Keeping it in the family

According to my assessment of current knowledge, of 253 recognised taxonomic families of birds, over one quarter (74 families) contain birds that visit flowers and are known or presumed pollinators. As I mentioned, that came as a surprise. Whilst I knew that nectarivory was going to extend far beyond those families that so readily spring to mind, such as the hummingbirds and sunbirds, I had no idea that it would encompass such an evolutionary diversity of birds. Based on what I said in the previous chapter, this tells us that nectar feeding has evolved many times in the class Aves. This includes some recently extinct species that we know from museum specimens were regular flower visitors. We'll return to these in Chapter 18.

In order to appreciate this family-level diversity, I've laid out the different groups of birds that are known to be flower visitors in a table on pages 29–32.

There are a few different ways of ordering a table like this. I could do it alphabetically, from A (Acanthisittidae – the New Zealand wrens) to Z (Zosteropidae – the white-eyes, yuhinas and their relatives), but that's purely arbitrary and has no real scientific meaning. It's library-book biodiversity. I could also sort by the number of species in each group, from the most diverse to the least. But would I start with hummingbirds (Trochilidae), a family that contains 363 species, all of them flower visitors and potential pollinators? Or with the tyrant flycatchers (Tyrannidae), which are more diverse, with 425 described species, though only a small proportion of them are recorded as flower visitors? Either way, it would be a little too hierarchical for my taste, though it would tell us a lot about the ecological dominance and evolutionary diversification of these different groups with respect to their potential abilities as pollinators.

A better way to arrange the table might be biogeographically, so that it represents the diversity of flower-visiting birds in different regions. But that presents us with the problem of where to place those families that are found over large, and biogeographically disparate, regions – for example the Old World parrots (Psittaculidae), which range from sub-Saharan Africa to Asia, Australasia and on to the Pacific islands. Or the finches, euphonias and allies (Fringillidae), which are globally distributed except for Southeast Asia and Australasia.

These are the kinds of considerations that exercise the minds of all scientists: how best to order data, whether in tables or in graphs and other kinds of figures. And the order certainly does matter, because there are all kinds of hidden messages in the way in which data are sorted. For example, I did not want to begin the table with the hummingbirds at the top, even though it's a logical place to start from the point of view of diversity, because these birds dominate the scientific literature, natural history documentaries, and our wider consciousness of what constitutes a 'pollinating bird'. One of the purposes of this book is to highlight the 'other' bird pollinators and, frankly, the hummingbirds grab more than their share of the limelight.

After some consideration I decided to arrange the table phylogenetically, so that those bird families that are most closely related are grouped together in their respective orders. I've started with the order with the most families, the passerines (Passeriformes), because I think this gives a sense of the ecological importance and evolutionary dominance of this group of birds compared with the others. Within each order, the families are arranged from the most to the least diverse.

Remember that this table is bound to be incomplete, because there will be published information that I have missed and new observations still to be made. The ecologies of many of these birds as flower visitors are not well studied, and there is still much to be discovered even by 'amateur' naturalists and 'citizen scientists'.

The diversity of flower-visiting bird orders and families

Informal name	Family	No. of species	No. known to visit flowers	Distribution
Order Passeriformes (passerines, 'perching birds', 'songbirds')				
Tyrant flycatchers	Tyrannidae	425	13	Americas
Tanagers, honeycreepers, Galápagos finches and allies	Thraupidae	381	86	Central and South America, Galápagos
Old World flycatchers	Muscicapidae	327	5	Eurasia, Africa, Southeast Asia, Arctic North America
Ovenbirds and woodcreepers	Furnariidae	306	1	Central and South America
Finches, euphonias, Hawaiian honeycreepers, and allies	Fringillidae	229	26	Global except Southeast Asia and Australasia
Honeyeaters	Meliphagidae	189	189?	Australasia and Pacific Islands
Thrushes and allies	Turdidae	174	6	Global
Cisticolas and allies	Cisticolidae	161	6	Southern Europe, Africa, Middle East, Asia, Australasia
Bulbuls	Pycnonotidae	151	8	Old World tropics and subtropics
Sunbirds and spiderhunters	Nectariniidae	143	143?	Old World tropics and subtropics, into Australasia
Laughingthrushes and allies	Leiothrichidae	143	1	Old World tropics and subtropics
White-eyes, yuhinas and allies	Zosteropidae	142	142?	Sub-Saharan Africa, Asia, Australasia, Pacific islands
Waxbills and allies	Estrildidae	140	5	Sub-Saharan Africa, Middle East, Asia, Australasia
New World sparrows	Passerellidae	132	1	Americas
Crows, jays and magpies	Corvidae	128	16	Global
Starlings	Sturnidae	123	12	Old World
Weavers and allies	Ploceidae	118	7	Old World tropics and subtropics
New World warblers	Parulidae	111	18	Americas
Troupials and allies	Icteridae	105	48	Americas
Wrens	Troglodytidae	86	5	Eurasia and Americas

(continued)

The diversity of flower-visiting bird orders and families (*continued*)

Informal name	Family	No. of species	No. known to visit flowers	Distribution
Leaf warblers	Phylloscopidae	79	7	Old World
Sylviid warblers, parrotbills and allies	Sylviidae	69	7	Old World
Wagtails and pipits	Motacillidae	67	1	Global
Thornbills and allies	Acanthizidae	66	3	Australasia, southern Asia and Indonesia
Tits, chickadees and titmice	Paridae	63	4	Northern hemisphere and Old World tropics
Vireos, shrike-babblers and *Erpornis*	Vireonidae	63	4	Americas and southern Asia
Reed warblers and allies	Acrocephalidae	59	1	Old World
Tree-babblers, scimitar-babblers and allies	Timaliidae	58	3	Southern Asia
Fantails	Rhipiduridae	54	1	South Asia and Australasia
Bushshrikes and allies	Malaconotidae	50	1	Sub-Saharan Africa
Cardinals and allies	Cardinalidae	49	6	Americas
Flowerpeckers	Dicaeidae	50	50	Tropical southern Asia and Australasia
Old World buntings	Emberizidae	44	6	Eurasia and Africa
Old World sparrows	Passeridae	43	1	Eurasia, Africa, Southeast Asia
Old World orioles	Oriolidae	38	1	Eurasia, sub-Saharan Africa, Southeast Asia, Indonesia, Australasia
Shrikes	Laniidae	34	1	North America, Eurasia, Africa, Asia, Southeast Asia
Mockingbirds and thrashers	Mimidae	34	8	Americas
Tityras and allies	Tityridae	33	1	Neotropics
Drongos	Dicruridae	29	1	Old World Tropics and Subtropics
African warblers	Macrosphenidae	21	2	Sub-Saharan Africa
Accentors	Prunellidae	12	1	Eurasia, North Africa
Long-tailed tits	Aegithalidae	11	1	Eurasia and western North America

Informal name	Family	No. of species	No. known to visit flowers	Distribution
Leafbirds	Chloropseidae	11	11?	Indian subcontinent and Southeast Asia
Berrypeckers and longbills	Melanocharitidae	11	11?	New Guinea
Penduline-tits	Remizidae	11	2	Eurasia, sub-Saharan Africa and western North America
Sharpbill, Royal Flycatcher and allies	Oxyruncidae	7	1	Neotropics
Kinglets	Regulidae	6	1	North America and Eurasia
Wattlebirds	Callaeidae	5	2	New Zealand
Hawaiian honeyeaters	Mohoidae	5	5	Hawaii
Hispaniolan tanagers	Phaenicophilidae	4	1	Endemic to Hispaniola
Asities	Philepittidae	4	4	Madagascar
Spindalises	Spindalidae	4	1	Caribbean
New Zealand Wrens	Acanthisittidae	4	1	New Zealand
Whiteheads	Mohouidae	3	2	New Zealand
Fairy-bluebirds	Irenidae	2	1	Indian subcontinent and Southeast Asia
Painted berrypeckers	Paramythiidae	2	2	New Guinea
Sugarbirds	Promeropidae	2	2	Southern Africa
Yellow-breasted Chat	Icteriidae	1	1	North and Central America
Stitchbird	Notiomystidae	1	1	New Zealand
Order Apodiformes (swifts, treeswifts and hummingbirds)				
Hummingbirds	Trochilidae	363	363	Americas
Order Bucerotiformes (hornbills, hoopooes, etc.)				
Woodhoopoes and scimitarbills	Phoeniculidae	8	2	Sub-Saharan Africa
Order Charadriiformes (gulls and shorebirds)				
Seedsnipes	Thinocoridae	4	1	Southern South America
Order Coliiformes (mousebirds)				
Mousebirds	Coliidae	6	2	Sub-Saharan Africa
Order Columbiformes (pigeons and doves)				
Pigeons and doves	Columbidae	348	9	Global

(*continued*)

The diversity of flower-visiting bird orders and families (*continued*)

Informal name	Family	No. of species	No. known to visit flowers	Distribution
Order Cuculiformes (cuckoos)				
Cuckoos	Cuculidae	147	1	Global
Order Galliformes (landfowl, quails, megapodes, etc.)				
Guans, chachalacas and curassows	Cracidae	56	1	Neotropics
Order Piciformes (woodpeckers and relatives)				
Woodpeckers	Picidae	233	18	Global except Australasia
African barbets	Lybiidae	41	3	Sub-Saharan Africa
Honeyguides	Indicatoridae	17	1	Sub-Saharan Africa
Order Psittaciformes (parrots)				
Old World parrots	Psittaculidae	190	75	Sub-Saharan Africa, Asia, Australasia, Pacific islands
New World and African parrots	Psittacidae	175	21	Central and South America, sub-Saharan Africa
Cockatoos	Cacatuidae	21	3	Australasia
New Zealand parrots	Strigopidae	4	1	New Zealand
Order Trogoniformes (trogons and quetzals)				
Trogons and quetzals	Trogonidae	43	2	Global tropics

In each case, the number of species known to visit flowers should be considered a minimum estimate. The number of species in each family is based on accounts in the Cornell Lab of Ornithology's *Birds of the World*. Distributions refer to the whole family, not just the flower visitors, and are categorised very broadly.

In September 2023, just as I thought I'd completed the final edits for this book, Australian ecologist Stewart Nicol emailed me about a recent study he had published on an endemic Tasmanian plant called *Richea dracophylla*. This enigmatic member of the heather and rhododendron family (Ericaceae) is known to be pollinated by honeyeaters and insects. Stewart sent a video of a Black Currawong (another endemic) feasting on the flowers but also, potentially, pollinating some with its feet (see Chapter 6). Currawongs belong to the family Artamidae, the woodswallows, bellmagpies and allies. Further research is required, but this family may need to be added to the table, increasing the (current) total to 75 families. So we proceed.

Science is not a set of fixed facts, it is always developing as new knowledge becomes available. Nonetheless, this table is a good starting point for understanding the true diversity of the birds that visit, and often pollinate, flowers.

In the remainder of this chapter we'll explore this table and introduce the main groups of birds in turn, though of necessity this will be brief. We'll encounter most of these groups in later chapters and of course there are books devoted to some of them. In addition, there is a collection of bird families that include only one or two species that have been observed as flower visitors; I'll pay less attention to these, with the proviso that they may be important pollinators of some plants under some circumstances. In ecology, context is everything.

Perching birds – Passeriformes

Collectively, this order of birds includes 60% of the total number of living bird species, classified into at least 140 families. Over 40% of these, 59 families, have records of flower feeding, in some cases just a single species within a larger family, such as the accentors (Prunellidae), the New World Sparrows (Passerellidae) and the wagtails and pipits (Motacillidae). However, the order also includes families for which we know or strongly suspect that flower feeding is a vital aspect of the lives of all or most species and which are regionally important pollinators, such as the honeyeaters (Meliphagidae), the white-eyes, yuhinas and allies (Zosteropidae) and the sunbirds and spiderhunters (Nectariniidae). These highly specialist nectar-feeding groups we will discuss in some detail throughout the book, so for now I want to focus on some lesser-known families that often surprise us with their nectarivorous habits. Having said that, it shouldn't come as a shock that some of the crows and their relatives (Corvidae) are known to visit flowers, including at least a quarter of the 40-plus species of the genus *Corvus*, according to a recent review by James Fitzsimons. In my experience, these highly intelligent and generally omnivorous birds will investigate anything that even hints at being a source of food.

The less familiar pollinator families include some in which a significant proportion of the species are known to be flower visitors. Amongst these are the troupials and allies (Icteridae), a group that's found throughout the Americas. Of the 105 species, almost half feed on flowers, most of which are species of oriole (*Icterus*). Not to be confused with the Old World orioles, which possess the same colloquial name but belong in a different family (Oriolidae).[1] These New World species are distinctively coloured birds that consume a wide range of invertebrates and fruit, as well as nectar.

Another significant family of pollinating passerines is the Thraupidae, which includes the tanagers, honeycreepers and Darwin's finches (whom we'll meet in Chapter 12). Almost a quarter of species in this large family have been observed visiting, or at least exploiting, flowers and it includes at least two genera which are flower specialists, though in very different ways. The flowerpiercers (*Diglossa* spp.) have distinctive hooks to the end of their bills that they use to rob nectar from flowers by creating holes in the base; this means that they miss the sexual parts of the flowers and are traditionally thought of as parasites of flowers rather than true pollinators. However, a 2021 study of flowerpiercer ecology in the high Andes of Colombia showed that we must revise this view: these birds can, and do, carry significant amounts of pollen of some flowers, showing that not all of their activities involve stealing nectar through the back door: they also enter flowers from the front. Not only that, but nectar robbing itself can sometimes have an unexpectedly positive outcome for a plant's reproduction, as we'll see later.

A further surprise awaits us when we consider the finches (Fringillidae), normally thought of as a predominantly granivorous (seed-eating) group of birds. Amongst the more typical finches, some are confirmed as pollinators, including such familiar birds as the Common Chaffinch and the Island Canary. However, this family also includes the nectar-specialist Hawaiian honeycreepers, which recent DNA analyses place as close relatives of the rose-finches. Many species within this group are sadly extinct, as we'll see in Chapter 18.

The final passerine families that I want to deal with provide a good illustration of the limits of our knowledge and the challenges

---

1 Also not to be confused with the family Icteriidae (double 'i'), which contains just one species, the Yellow-breasted Chat of North and Central America – also a known flower visitor. Taxonomists sometimes make odd and confusing choices when they give scientific names to groups.

that we face when it comes to documenting bird–flower interactions. Despite their informal name, the 48 species of flowerpeckers (Dicaeidae) were not included in the 'Global trends' paper by Eugenie Regan and colleagues (see Chapter 2), probably because there are few published studies of their feeding behaviour. Flowerpeckers are species of tropical rainforest canopies, a notoriously difficult habitat in which to perform detailed ecological and behavioural studies. Species in the largest genus in the family, *Dicaeum*, are furnished with fringed, tubular tongues similar to honeyeaters, well adapted for mopping and supping nectar, and the few species that have been studied to date do visit flowers. I've therefore included the whole family in the table.

Even less well known than the flowerpeckers are two small New Guinea families called the berrypeckers and longbills (Melanocharitidae), with 11 species, and the painted berrypeckers (Paramythiidae), with only two species, the Tit Berrypecker and the Crested Berrypecker. According to *Birds of the World*, the former includes flowers in its diet while the latter has 'been seen to pluck large ericaceous flowers and smear themselves with them', an odd and perhaps unique behaviour that could have an ornamental or medicinal function and deserves to be further studied. Another poorly documented group is the leafbirds (Chloropseidae). Based on scattered observations, these species are assumed to include significant amounts of nectar in their diet, and, perhaps more significantly, have adaptations such as the finely frayed tips of their tongues used to mop up nectar. Such flower-adapted tongues have occasionally evolved in species belonging to otherwise non-nectarivorous families, such as the Brahminy Starling. The Fire-tailed Myzornis likewise has a tongue well suited for nectar feeding and regularly visits flowers. This is one of the sylviid warblers (Sylviidae), which are predominantly insect and fruit eaters, though the family also includes species such as the Eurasian Blackcap and Sardinian Warbler, which are rare, but confirmed, pollinators in Europe (see Chapter 13).

Finally in the passerines, an intriguing small family is the asities (Philepittidae), which comprises only four species, all endemic to Madagascar. Two of these are occasional nectar feeders, whilst the other two rely heavily on flowers. With their brightly coloured feathers and long, downcurved beaks, the latter species look superficially like sunbirds and have appropriate common names: the Common Sunbird-Asity and the Yellow-bellied Sunbird-Asity.

Parrots and their relatives – Psittaciformes

Flamboyantly coloured parrots and parakeets are the mainstay of many a documentary about tropical birds. However, their distribution spans a much greater range than just tropical latitudes, extending into subtropical and warm regions of the Americas, Africa, Asia and Australasia. Not only that, but the order comprises four families that are separated by perhaps as much as 60 million years of evolution and thousands of kilometres: the New Zealand parrots (Strigopidae), the cockatoos (Cacatuidae), the New World and African parrots (Psittacidae) and the Old World parrots (Psittaculidae).

Although seeds and fruit are predominantly consumed by most parrot species, flower feeding and at least potentially pollinating has been documented in all four families. The New Zealand Kaka, one of only four species within that family, has a fringed tongue and includes a significant proportion of nectar (and indeed tree sap) in its diet. Flower feeding is most frequently encountered in the Old World parrots, however, especially Australian members of the family, and flowers can be extremely important in their diets during some periods of the year. This achieves its most specialised form in the lories and lorikeets (the subfamily Loriinae) where, again, the tips of the tongues are finely divided and more effectively take up nectar and pollen from flowers. Some endemic Australian plants such as species of banksia and eucalyptus have flower heads that are tough enough to survive the thuggish behaviour of these often quarrelsome birds. In suburban Sydney I've watched confiding Rainbow Lorikeets feeding on banksias, oblivious to passing cars and curious biologists, and moving a lot of pollen around between trees (Plate 4).

However, there's no doubt that, as well as being pollinators, many members of this order of birds destroy flowers that are not well adapted to their depredations, and some fruit farmers consider them pests. For the birds, the line between herbivore and pollinator is not one which they consider important: for them it's purely about gaining a source of food. How the flowers adapt to their behaviour is down to the plants.

Woodpeckers and barbets – Piciformes

Woodpeckers (Picidae) might appear to be unlikely candidates as pollinators as they are generally insectivores that probe in the soil, under bark and in dead wood for small invertebrates, or are predators of small vertebrates and eggs. Some species, however, also eat fruit and tree sap. Indeed, species in the North American genus *Sphyrapicus* have earned the informal name of sapsuckers because a high proportion of their adult diet is the sap of living trees.

The long, pointed beaks, and the even longer tongues, that woodpeckers use for wheedling bugs out of bark and sap out of stems are also useful tools for accessing nectar in deep flowers. There are no woodpeckers (that we know of) that are flower specialists. However, almost 10% of species are recorded as visiting flowers for nectar, and these elongated beaks and tongues are certainly helpful. In 2010 my Danish friend and colleague Bo Dalsgaard was carrying out field work in Cuba when he observed a Cuban Green Woodpecker probing the flowers of the Geiger Tree (Plate 5). In Bo's own words:

> The woodpecker was perched on the same or a nearby branch as the inflorescences, probing flowers above and adjacent, as well as inverting while probing flowers underneath. It inserted the bill into the floral tube then drank nectar for one to several seconds, touching anthers and stigma with the bill.

This acrobatic bird was almost certainly a pollinator, given that it was touching the sexual parts of the flame-red flowers of these native Cuban trees, which Bo found were visited by at least one other bird species, the Cuban Emerald hummingbird. The large amounts of highly dilute nectar produced by the narrow, trumpet-shaped flowers suggests that the tree is a bird specialist, though it probably attracts a range of birds. Reviewing what was then known about woodpecker pollination in the Neotropics, including reaching out to ornithologists for unpublished observations, Bo concluded that the phenomenon is almost certainly under-recorded, although some species have never been observed to visit flowers, despite extensive field observations of their diets.

I have to agree with Bo, and in fact I would go further: woodpecker pollination is undoubtedly under-recorded everywhere, including in Asia, where our observations of the Darjeeling Woodpecker visiting flowers of rhododendrons in Nepal was the first record for that biodiverse country (see Chapter 12). In North America, Nick Waser tells me that Gila Woodpeckers will often take sugar water from hummingbird feeders in his garden, aggressively displacing hummers and Verdins. To date, almost 20 species of woodpecker have been seen visiting flowers, and some are undoubtedly pollinators. Intriguingly, a 120-million-year-old fossil bird recently described from the Early Cretaceous of China is described as being

> the earliest example of a bird being able to stick its tongue out .... this feature makes one wonder why this bird would be sticking its tongue out. We hypothesize that [it] might have used this feature for catching insects in the same way that living woodpeckers use their tongues to get insects out of holes ... Alternatively, the bird might have been feeding on pollen or nectar-like liquids from plants in the forest where it lived.

If in the future that's proven to be correct I'll have to amend the subtitle of later editions of this book!

As well as the woodpeckers, from this order there are also a handful of barbets that are confirmed pollinators, plus intriguingly the Lesser Honeyguide of sub-Saharan Africa, whose interactions with bees and humans are well documented.

Hummingbirds and swifts – Apodiformes

It surprises many people to learn that the closest relatives of the hummingbirds (Trochilidae) are not the passerine sunbirds or the honeyeaters, which they resemble in so many ways, but rather the swifts (Apodidae) and the treeswifts (Hemiprocnidae). But these relationships have long been supported by studies of the birds' skeletons and anatomies, and confirmed more recently by DNA research. This relationship is also apparent when one views the flight of the Giant Hummingbird – which, according to *Birds of the World*, 'can be confused on the wing with a swallow or a swift, because of its large size and peculiar flight, in which short glides often are interspersed among the slow wingbeats.'

With the exception of the long-extinct species known only from fossils in Europe that I discussed in Chapter 1, all hummingbirds are found in the Americas and associated islands. They range from southern Alaska in the north, where the migratory Rufous Hummingbird can be found in the summer, to the tip of South America, from where firecrowns in the genus *Sephanoides* migrate north for the southern winter.[2] Hummingbirds are also especially diverse within the islands of the Caribbean, and have even found their way to more distant Pacific archipelagos such as the Juan Fernández group.

With 363 described species, the hummingbirds are almost twice as diverse as the next-largest specialist flower-visiting family, the honeyeaters. Traditionally this family was divided into two subfamilies, the 'hermits' and 'non-hermits'. However, recent phylogenetic analyses have shown that this binary division is artificial and it's better to consider them as a series of six subfamilies containing various tribes. As well as the hermits, these

2 The Rufous Hummingbird winters in Central America. It was apparently American biologist William A. Calder III who first pointed out that this is the longest known migration of any bird *when measured in bird body lengths.*

are given informal names that are descriptive and/or flamboyant as required – as exemplified by the topazes, mangoes, coquettes, emeralds and mountain gems.

Few other groups of birds, and certainly none of the other pollinating groups, have been studied as well or as intensively as the hummingbirds. As we'll discover in later chapters, we probably know more about their biology and ecology than all of the other bird pollinators combined. And although they are often thought of as the most specialised of the flower visitors, they cannot and do not subsist on nectar alone: all species are known to take insects, spiders and other invertebrates, and occasionally the juice of fruit and tree sap.

A final miscellany

The remaining orders in the table are a ragbag of groups in which visitation to, and pollination of, flowers has only rarely been observed – the woodhoopoes and scimitarbills (Bucerotiformes – Plate 6), seedsnipes (Charadriiformes), mousebirds (Coliiformes), guans, chachalacas, and curassows (Galliformes), cuckoos (Cuculiformes), and trogons (Trogoniformes). An exception here is perhaps seen in the pigeons and doves (Columbiformes), some of which can be regular and important pollinators of desert cacti. That iconic emblem of many a Western movie, the Saguaro cactus, for example, is pollinated by bats and moths at night and by bees and birds during the daytime, including Gila Woodpeckers, Costa's Hummingbirds and White-winged Doves. Both the bats and some of the birds also feed on the Saguaro's fruit, dispersing its seeds in the process, a rare example of what's termed a 'double mutualism' in which one species provides two distinct benefits to another organism.

Accumulating diversity

This chapter and the previous one present a state-of-the-art account of the families of birds, and an estimation of the number of species, that interact with and probably pollinate flowers. Unfortunately, the state of the art is constantly shifting and there are at least three sources of inaccuracy that will render the estimates out of date within a few years, if not sooner.

First, I will have missed literature about some of the more obscure genera and families that visit flowers. In fact I almost missed Truman Young's account of the Moorland Chat (called the 'Mountain Chat' in that paper) as a probable pollinator of Mount Kenya lobelias, which added a new family, the Old World flycatchers (Muscicapidae) to my database. I only found that paper when Alexander Schlatmann sent me the photograph shown in Plate 7. It's doubly embarrassing because I'd met Truman in Kenya in 2022 and he mentioned that he'd worked on these plants! But that's the thing with the scientific literature: it's so vast that it's very easy to miss relevant information. Every scientist has had the experience of reading a published account of research close to their own subject area and thinking: 'Why didn't they cite *my* paper about this!?' To all of those scientists, myself included, I say: remember that you're as likely to be a culprit as a victim.

The second source of inaccuracy stems from the fact that there are new discoveries waiting to be made about interactions between birds and flowers. This is certainly true in the biodiverse regions of the world, especially the tropics. For example, it wasn't until the early 2000s that we really understood that woodpeckers could be pollinators of flowers. But this can also occur with species that are common and familiar, as I will recount later when we consider 'the curious case of Europe' in Chapter 13.

Finally, we are still describing species new to science in many bird families. As I was drafting this chapter in late 2022, I spotted a report by a team from Trinity College Dublin and Universitas Halu Oleo in Sulawesi. Their genetic data confirm earlier studies suggesting that there is 'cryptic' diversity of sunbird species across Indonesia. Olive-backed Sunbirds should be split into several species, including one endemic to the Wakatobi Islands, whilst Black Sunbirds from Sulawesi are almost certainly a different, if closely related, species to those found in New Guinea. Further genetic studies of other groups of birds will no doubt produce similar findings.

An important question, of course, is how important some of these 'non-specialist' nectar-feeding birds are as pollinators. This is a topic that I'll return to in some of the later chapters. But it's worth stating now that for a lot of the interactions between birds and flowers we simply have not done the detailed experiments that are required to understand just how effective particular bird species are at depositing pollen on the stigmas of specific plants. However, in the cases where these experiments have been done, we can see that various tits, warblers and so forth can be extremely effective pollinators, even if they are only rare visitors to flowers. There are still huge gaps in our knowledge on this topic, however, including the true diversity of birds that rely on flowers and the range of flower types that they use. Fortunately, there are projects in progress that aim to answer such questions. For example, in 2022 as part of her PhD work, researcher Carolyn Coyle at Colorado State University set up the 'Songbirds as Pollinators' (SaP) project that will mobilise bird ringers ('banders' in American parlance) to swab the heads of songbirds that they have captured in order to discover what pollen is adhering to the feathers and bill. There's also the opportunity for others to submit records of flower feeding by these birds, so it's a truly inclusive 'citizen science' project. In this way Carolyn and the team hope to more fully understand the diversity of different flower types being used by these birds, especially during their annual migrations.

But all of this, from missing relevant research to new findings that debunk old ideas, is part and parcel of the process of science and of being a scientist, a recurring theme of this book. What I have presented here isn't wrong, it's just an assessment of the current state of our knowledge. And one thing is perfectly clear – on dozens of occasions in the distant past, birds have evolved behaviours and anatomies that have allowed them to exploit flowers. What does this mean for the evolution of the flowers that they visit? It's time to leave the birds behind for a while, and think about it from a plant's perspective.

CHAPTER 4

A flower's point of view

In the previous chapter I presented some robust figures on the number of species and families of birds that visit flowers and are certainly or potentially pollinators. It was possible to do that because birds are well studied and not especially diverse. Avian taxonomists disagree on what constitutes a 'species', and birds new to science are discovered every year, but as I noted previously, the general consensus amongst ornithologists is that there are around 11,000 living and recently extinct bird species. Even if we accept the radical notion that the true figure is twice that number, the scale of bird diversity is minuscule compared to insects, which are counted in the millions of species. Even fish are more diverse, with in excess of 34,000 species known, and no doubt many yet undiscovered from the deepest oceans of the world.

What about the plants? Again, botanists vary in their opinions, taxonomic disagreements abound, and new discoveries mean that the true diversity of plants is constantly being revised. But the world certainly supports not fewer than 300,000 species of flowering plants (the angiosperms) and the true figure could be as high as 400,000 species. Put another way, and depending which numbers we use in our calculation, for every bird species there may be between 15 and 40 different types of flowering plants. Most of these flowers are pollinated by insects such as bees, moths, flies and beetles, rather than by birds. But a significant fraction has co-opted birds to move their pollen around, as either their sole pollinators or in concert with other animals such as insects, bats or lizards.

As we'll see in the next chapter, sometimes the roles of birds are discounted because they don't fit with the expectations of what the flowers 'should' be pollinated by. Exploring the diversity of plants with flowers that are bird-pollinated is therefore a more difficult task than asking the question 'how many birds visit flowers?' There are two other reasons why it's more complicated. First of all, as noted, the flowering plants are about 30 times more diverse than the birds, so there is an issue about the scale of the task. More significantly, though, our knowledge of the interactions between flowers and their pollinators is not as complete as we might imagine. For my previous book *Pollinators & Pollination* I did a back-of-the-envelope calculation which suggested that we have data, of some kind, on the pollinators of about 10% of the world's angiosperms. I came to that conclusion by looking at reviews of pollination in plant families and in particular regions of the world and assessing how complete these surveys were. For example, a synthesis of what's known about the pollinators in the huge dogbane and milkweed family (Apocynaceae[1]) uncovered data for about 12% of the more than 5,300 species in the family. More about this, my very favourite plant family, later.

Research published since then appears to support the guesstimate in my book. For example, in 2022 Chilean ecologists Giselle Muschett and Francisco Fontúrbel published 'A comprehensive catalogue of plant–pollinator interactions for Chile', containing data on the flower visitors to 357 species of plants belonging to 83 angiosperm families. Now, Chile is a plant diversity hotspot for a number of reasons, not least that it is long and thin and extends from the tropics in the north to cold temperate regions in the south, and spans elevations from sea level to more than 6,800 metres in the Andes. Not only that, but Chile as a political entity includes some oceanic islands which, as we will discuss in Chapter 12, often support endemic species, i.e. those found nowhere else in the world. This geographical and topographical diversity means that the flora of Chile amounts to somewhere between 4,000 and

1 In Chapter 2 I noted that animal family names all end in –idae. Just as helpfully, plant family names end in –aceae, except for a very few families when written by some die-hard botanical traditionalists who insist, for example, that the daisy family is the Compositae rather than the Asteraceae.

5,000 vascular plant species.[2] The vast majority of these are flowering plants, but even if we exclude groups such as ferns and conifers, and use a conservative figure of 3,500 species of flowering plants, and note that a proportion of the plants in their database are not native to Chile, Giselle and Francisco's survey still amounts to a sample of just over 10%, in line with my previous calculation.

Historically there has been a tradition of active research on plant–pollinator interactions in Chile, and that continues to the present day. But given the size of the flora, and the scale of the task, collecting data on the flower visitors in this region is slow and incremental. Many regions of the world are even less well studied than Chile, and I would not be surprised if even my suggestion that we have data on the pollinators of 10% of the world's angiosperms is too high.

What all of this means is that any attempt to quantify the biodiversity of bird–flower interactions is going to be preliminary at this stage. But that doesn't mean that we shouldn't do it: all scientific enquiry is provisional and subject to possible revision in the future.

Assessing the bird flowers

How do we begin to quantify the number of bird-pollinated flowers that exist in the world? As a starting point, let's think about the considerable number of published research studies where pollination ecologists have surveyed an area of grassland, scrub or forest, and catalogued the plant species according to what pollinates them. Over the years I've collected these sorts of studies as I encounter them, and they provide a useful first pass at understanding, for example, how common bird pollination is compared to bee or fly pollination. It's worth exploring one or two examples of this sort of study.

In the world of pollination ecology, most people have heard of the Gottsbergers, the collective handle for German wife-and-husband

2 Vascular plants are those with a complex system of 'pipework' that conducts water and nutrients to leaves, stems and roots. As opposed to mosses, liverworts and so forth, which have no such conductive tissues.

team of Ilse Silberbauer-Gottsberger and Gerhard Gottsberger. They have worked in the field for decades, principally in Brazil, where they have made important and ground-breaking discoveries, especially in the Cerrado – a rather ill-defined but important biome that is increasingly being cleared for agriculture. Much less well known than Brazilian rainforest, this type of vegetation deserves wider attention and protection, because it's extraordinarily rich in species. In just one hectare of Cerrado that was studied intensively by the Gottsbergers, they identified 301 plant species. If these plants are classified according to their pollinators, we find that bird-pollinated flowers account for only five of them (less than 2% of the total). This is on the low side compared to other parts of the Americas: in both the equally well-researched Costa Rican rainforest at La Selva Biological Station and the Brazilian Caatinga (a type of tropical dry forest), around 15% of the plants are hummingbird-pollinated.

When I pulled together studies such as this for a review paper that I published in 2017, I concluded that on average across the world, 2.9% of plants within a habitat are solely bird-pollinated, with another 3% or so pollinated by both birds and insects. There's a lot of variation around both those figures, and individual surveys have found anywhere between zero and 26% of plants to be bird-pollinated. But let's add together those first two average estimates of bird pollination and round the total up to 6%. Almost 90% of flowering plants (say 315,000 species) are animal-pollinated, so 6% of those is 18,900. In other words, almost 20,000 plant species are likely to be wholly or in part bird-pollinated.

How does that figure compare with other published estimates? That's quite an easy question to answer because, as far as I'm aware, no one has tried to suggest a solid figure. All of the reviews of the topic that I've read say that 'thousands' of plants are likely to be bird-pollinated, which is certainly true. But is it low thousands? High thousands? Tens of thousands, as I'm suggesting? I would err on the higher side because there's at least one other study that points in such a direction. In 2015, Stefan Abrahamczyk and Michael Kessler published a paper entitled 'Morphological and behavioural adaptations to feed on nectar: how feeding ecology determines the diversity and composition of hummingbird assemblages', in which they suggested that around 7,000 plant species 'between Alaska and Tierra del Fuego … are dependent on hummingbirds for pollination'. That's a bold claim, but if it's correct, then the number of plants that are pollinated by sunbirds, honeyeaters, white eyes, generalist passerines and so forth elsewhere in the world must be at least as great.

Why is that? Well, Stefan and Michael's figure of 7,000 is based on the fact that, in the Americas, there are plant families that have a high proportion of species, indeed whole genera, that are hummingbird-pollinated. This includes big groups with thousands of species such as the acanthus family (Acanthaceae), the daisies (Asteraceae), the pineapple family (Bromeliaceae), the cacti (Cactaceae), the gesneriads (Gesneriaceae) and many, many more. If we move out of the realm of the hummingbirds, we find similar patterns for large plant families pollinated by other groups of birds. In Africa we encounter bird-pollinated balsams (Balsaminaceae), aloes and red hot pokers (Asphodelaceae), and members of the iris family (Iridaceae). Furthermore, the family Proteaceae has bird-pollinated species both in Africa (proteas) and in Australia (banksias and grevilleas), whilst groups such as the eucalypts (Myrtaceae) and many endemic genera in Australia are also bird-pollinated. A review by Greg Keighery in 1980 concluded that a minimum of 560 species of flowering plants are pollinated by birds, representing about 15% of the total southwestern Australian flora – which 'indicates adaptation to bird pollination is a major evolutionary force in this region'. Likewise, Richard Corlett's assessment of flower visitors and pollination in the Indomalayan region of Asia documented widespread examples of plants that are bird-pollinated, though he raised the interesting point that compared to pollination by small insects, the 'relative importance of bird pollination has certainly been exaggerated by its conspicuousness'. He may well be right, for pollination ecologists are often drawn to large, distinctive flowers for their studies – though, as we'll see in the next chapter, not all bird-pollinated flowers are like this.

I could go on documenting regions and plant families and genera with a significant proportion of bird-pollinated species, but I'm sure you get the point: avian pollination is widespread, ecologically important over much of the world, and has evolved many times in different groups of plants. I think that it's safe to say, however, that my initial estimate of 20,000 bird-pollinated flowers is likely to be fairly close to the mark. Before we look at some of these in detail, however, I want to clear up a terminological confusion.

Ornithophily *versus* ornithophily

Anyone who reads about bird pollination will encounter the words 'ornithophily' and 'ornithophilous'. They were invented in the nineteenth century and have been widely used since then. What's not generally acknowledged is that they are used in two distinct ways with meanings that only partially converge. This is problematic, because it can lead to confusion and misunderstandings.

The first meaning refers to the act of pollination by birds, as in 'the flower was ornithophilous' or 'in this plant genus, ornithophily is common'. This usage is purely ecological and refers to the interaction between these species.

The second way the words are used refers to the set of flower traits associated with classical ideas about bird pollination. Statements such as 'the flowers show all the hallmarks of ornithophily' or 'these flowers possess ornithophilous traits' occur regularly in the bird pollination literature.

Both of these distinct usages are acceptable, but they are not synonymous, such that a plant can be ornithophilous without being ornithophilous. Confused? Let me give you some examples. In this book I'll discuss flowers that are bird-pollinated (ornithophily definition 1) but which do not have the typical traits that have often been associated with bird pollination, such as reddish or orange tubular unscented flowers, large amounts of dilute nectar, and so forth. In contrast, I will also present accounts of plant species, from North America and South Africa, that do possess those floral features (ornithophily definition 2), but in which birds are either junior pollinators or are not involved in their pollination at all.

In other words, flowers can be ornithophilous in the first definition without being ornithophilous in the second definition, and vice versa. To quote a recent study by Cameroonian ecologist Onella Mundi and colleagues, for some species at least, 'ornithophily … does not guarantee a preference by sunbirds'.

Pollination by birds is not the only one of these 'pollination syndromes' to be so afflicted with dual meanings; it applies to all of them. I'll discuss the syndromes in more detail in Chapter 5; suffice to say here that it's incumbent on us researchers to be clear about our usage of words, but sometimes we are not explicit, and we don't define our terms. That's problematic in the case of syndromes because it leads to conflation: we begin to associate one use of the word with the other. The current Wikipedia entry for ornithophily, for example, begins as follows: 'Ornithophily or bird pollination is the pollination of flowering plants by birds … The association involves several distinctive plant adaptations forming a "pollination syndrome". The plants typically have colourful, often red, flowers with long tubular structures holding ample nectar and orientations of the stamen and stigma that ensure contact with the pollinator.'

Wikipedia is easily edited (perhaps a reader of this book can take on the task?), but there are similar statements in the published literature which help to keep this duality, and the potential confusion, alive. My solution in this book is to keep the use of 'ornithophily' to a minimum and simply refer to 'bird pollination'.

Now, back to the plants.

The spiny friends of birds

Are there any plants more iconic than cactuses? Their spiny, inflated forms, sometimes tall and elegant, often short and squat, are instantly recognisable to most people, though Cactaceae are sometimes confused with similar succulent species in other families. As I've mentioned, the Saguaro is possibly the most iconic cactus of all, and is at least partly pollinated by birds such as Gila Woodpeckers, White-winged Doves and Costa's Hummingbirds. But the relationships between cacti and pollinating birds do not end there.

Back in 2018 Pablo Gorostiague from the Universidad Nacional de Salta in Argentina came over to the UK to spend time with my research group. The main purpose of his visit was to collate and review what was known up to that point about the ecology and evolution of pollination systems in the cactus family. By 'pollination system' I'm referring to the combination of flower characteristics, such as colour, shape and scent, plus the identity of the pollinators at a fairly broad taxonomic level – for example, birds, bees or wasps. (A pollination *system* thus differs from a pollination *syndrome*, which is concerned only with the characteristics of the flower itself.) Although we ran into some issues that delayed publication of the study, including our respective work pressures and the COVID-19 pandemic, in 2022 our paper, with cactus expert Pablo Ortega-Baes as co-author, finally appeared.

By scouring the literature and adding in his own studies, Pablo was able to find information available on the pollinators of 148 cactus species, or a little over 10% of the family (there's that 'about 10%' figure again!). The most frequently encountered group of pollinators in the Cactaceae are the bees, which service just over 80% of the species. However, to our surprise, it turned out that birds were pollinators for almost one-third of the species. Sometimes the birds pollinated a cactus alongside bees or even bats, but in about 8% of species they were the sole service provider. Flowers adapted for bird pollination have evolved independently in at least five of the 16 recognised tribes of cacti and are mainly found in the tropical regions of the southern hemisphere. The flowers varied enormously in size, shape and colour. Some were long and tubular in form, and bright red, fitting the expectations of the hummingbird 'pollination syndrome' (see the next chapter). Others, however, were funnel- or bowl-shaped, and variously white, yellow, orange or pink. A formal analysis of these flower traits showed that they overlapped to some extent with flowers that were pollinated by other animals, especially bees. Most of the avian pollinators were hummingbirds, not surprisingly given that almost all cactus species are native to the Americas, but some passerines were also involved. As well as the White-winged Dove already mentioned, other work by Pablo and Pablo has shown that the Gray-hooded

Sierra Finch pollinates the endemic Argentinian cactus *Echinopsis leucantha* (Plate 8).

Birds weave their way through bees and moths

Less immediately familiar than the cacti, but larger in terms of number of species and geographical distribution, is the dogbane and milkweed family (Apocynaceae). It's about the tenth biggest plant family and most of its species produce a white latex, hence the name 'milkweed'. Species of *Asclepias* are the caterpillar food plants of the well-known Monarch butterfly, plants such as hoyas and mandevillas are often grown on windowsills, whilst periwinkles and oleanders are familiar garden plants in temperate and subtropical areas of the world. When we reviewed the diversity of pollination systems in the family we found that although birds have been quite often recorded as flower visitors, confirmed bird pollination was extremely rare. I suspect that these plants actually use birds as pollen vectors more often than we currently know. Certainly, as with the cacti, the interactions between birds and flowers have appeared multiple times in different tribes within the family. We know this because we can use differences in DNA or RNA to reconstruct the evolutionary history of a group of organisms.

The evolutionary history of a group of plants such as the dogbanes and milkweeds, at whatever taxonomic level, from order to family to genus, can be viewed as a 'tree', with each branch corresponding to a particular plant lineage. The points where the branches fork represent times at which lineages have split. Perhaps two families or genera have begun to diverge, or two species have started on different evolutionary trajectories. If we just consider a species-level tree, then sometimes those diverging branches merge back together, much as the adjacent branches of a tree can fuse if they have been in contact with one another for long enough. Such hybridisation is not uncommon in the flowering plants and can give rise to new species.

Evolutionary or 'phylogenetic' trees such as this are now routinely constructed using data from DNA or RNA sequencing, and this allows a fairly objective view of the phylogeny of a group of species, in other words their relatedness in terms of evolution. Species that are on adjacent branches of the tree are more closely related than those that are more distant, in the same way that siblings are more closely related than cousins in human families. These sorts of evolutionary trees can also be constructed using other biological characteristics related to morphology and anatomy, and indeed this was the only way to do it prior to the development of molecular approaches to understanding biodiversity.

In a species-level tree, if we follow the branches to their tips we find the plants as they are now. At that point we can start to hang species' characteristics onto the phylogeny, much as we'd dress up a Christmas tree with baubles. These characteristics could be things like seed size or leaf shape, growth habit or biogeographic region. By understanding how the characteristics of different species differ or are the same, we can begin to infer how these species evolved. Pollination systems – the interplay between flower characteristics and the pollinator(s) of that plant – are especially interesting ornaments with which to adorn the tree, because they may reflect the defining characteristics of the plant species if they have evolved due to natural selection from different types of pollinators.

There is an increasing number of studies that have mapped pollination systems onto species' phylogenies, and some of these include plants that are bird-pollinated. This sort of research is telling us some very interesting things about how often, when and where bird-pollinated flowers evolve from those that are pollinated by other kinds of animals. Or indeed when pollination by animals such as bees and moths has evolved in species with bird-pollinated ancestors. Let me describe some examples.

In the early 2000s, a team of scientists including Maria Clara 'Cala' Castellanos and Paul Wilson started to piece together the story of how bee and bird pollination switched back and forth during the evolution of the beardtongues or penstemons. This is a group of plants that includes not only the genus *Penstemon* but also some closely related genera such as *Keckiella*. They were formerly included in the figwort family (Scrophulariaceae), though recent molecular phylogenetic analyses of the type described above have shown that they are actually plantains (Plantaginaceae).[3] The penstemons include somewhere between 270 and 300 species, most of which are bee-pollinated. In the studies led by Cala and Paul, it was estimated that perhaps 39 of those species are either partly or completely pollinated by hummingbirds. When viewed phylogenetically, it was clear that some degree of hummingbird pollination had evolved from exclusive bee pollination on a minimum of 13, and possibly as many as 25, occasions during the past history of this group of plants.

These switches from bee to bird pollination correlate with changes in the flowers. For example, bee-pollinated flowers tend to be purple, bluish or yellow in colour, whereas penstemon flowers that use birds are usually red, orange, magenta or rose-coloured to human eyes. There are also differences in the nectar, with 'bird flowers' typically having larger amounts of more dilute nectar than 'bee flowers'. In addition, flowers that are bird-pollinated often have sexual parts that extend further beyond the mouth of the flower tube and lack a large 'lip' on which bees can land.

If we take the extremes of the number of times hummingbirds have been recruited (25) and a conservative estimate of the number of species of penstemon (250), then we see that there has been a shift to partial or exclusive bird pollination in at most only 10% of the evolutionary events in this group of plants. All of the other cases of evolution of new species have been rather conservative in terms of recruiting new pollinators; in other words, bee-pollinated ancestors usually give rise to bee-pollinated descendants.

This sort of detailed analysis has only been done for a handful of plant groups, but in most cases similar patterns emerge. In 1965 (the year I was born, incidentally) husband and wife team Verne and Karen Grant produced the book *Flower Pollination in the Phlox Family*. They brought together their published and unpublished observations to document the range of different pollinators

---

3 Completely unrelated to the plantains that are green bananas; they belong in the family Musaceae. This is why we use scientific names for organisms.

that are exploited by these plants, including bees, butterflies, moths and hummingbirds. Since then, this small family (Polemoniaceae, with only 400 or so species) has become a model system for understanding how plants diversify. In the most recent analysis of the family, Jeffrey Rose and Kenneth Sytsma discovered that, as with the penstemons, hummingbird pollination is nowhere near as common in the family as bee or moth and butterfly pollination, but it has arisen independently on at least five occasions. Moreover, there seems to be much greater lability and switching between pollinators than was observed in the penstemons, with bird pollination evolving from ancestors that had been pollinated by bees, moths and butterflies, and long-tongued flies, as well as some that were self-pollinated. Conversely, bird-pollinated ancestors were shown to have evolved into modern descendants which use all of those pollinators, as well as bats.

All of these examples, and others that there's not space to describe, raise an important question that I hinted at earlier in the book.

Why birds?

Compared to birds, insects such as bees, flies and butterflies are far more diverse and abundant, and therefore more reliable as pollinating partners. So the question of why any plant should give up on insects and evolve a pollination relationship with birds is an important one to try to understand. One factor to consider is that birds can potentially move the genes of plants over a much wider area than most insects. Ecologists Diana Gamba and Nathan Muchhala, working in the Andean cloud forests of Ecuador, studied pairs of plant species within the same family, one insect- and one hummingbird-pollinated. They found that most of the time populations of the hummingbird-pollinated species were genetically more similar to one another. This implies greater 'gene flow' between the plants that use birds to carry their pollen, and points to birds as providing more effective pollination services in those plants that are naturally rather rare and scattered across a landscape, as is common in the tropics. Greater gene flow doesn't happen in all cases, because as Diana and Nathan point out, different bird behaviours – particularly whether or not the birds are territorial and have restricted movements – can affect how pollen moves in a population.

There is something of a paradox here, however. If birds are so successful at maintaining the genetic integrity of scattered populations of plants, why is it that, in some plant groups at least, bird pollination is also associated with an increase in the rate of species formation? As Martha Serrano-Serrano and co-authors note in their study of Neotropical species in the Gesneriaceae:

> Plants with a hummingbird pollination syndrome have a twofold increase in the rate of speciation, suggesting a positive effect of hummingbird pollination on the establishment of reproductive isolation.

Reproductive isolation has long been considered an important factor in the formation of new species, but it requires that there is limited movement of genes (via pollen or seeds) between populations, which is the reverse of what Diana and Nathan are suggesting is the advantage of adopting birds as pollinators. This contradiction is just one of many still unanswered questions about birds as pollinators.

Whether a plant community has a high or a low diversity of bird-pollinated plants within it depends on many interacting factors, not least the diversity of nectar- or pollen-feeding birds that is found in the surrounding region. There are other, more local factors that are important too. For example, in the study described above, the Gottsbergers explain the low frequency of bird flowers that they encountered as being due to hummingbirds finding a more reliable and continuous supply of nectar in adjacent gallery forests that line river courses. Conversely, in the Cerrado, most plants stop flowering in the dry season. They also note that hummingbirds can be opportunistic (as they see it) and take nectar from flowers that are not 'ornithophilous' in appearance. Many researchers have taken this more traditional approach and have assumed that, from the plant's perspective, only those flowers that show a certain limited set of flower traits are adapted to bird pollination. As we shall see in the next chapter, this does impose some limitations on the conclusions that we can draw about how important birds are for plants, and plants for birds.

CHAPTER 5

In the eye of the beholder

During my years of university teaching, I've always been a firm believer in the value of hands-on education such as field or laboratory work. One of practical exercises I used to run involved asking the students to take samples of nectar from plants around the university campus. Using fine, calibrated glass tubes to draw out the nectar, they would first measure the volume and then squirt the nectar onto the prismatic surface of a piece of equipment called a sugar refractometer. This measured the proportion of sugar in the fluid. Both of these are standard techniques employed by pollination ecologists in the field that provide a reasonable assessment of the nectar characteristics of flowers, though much more sophisticated approaches can be employed in a well-equipped laboratory. I also asked the students to make notes about the size and shape and positioning of the flowers. Back in the laboratory, the students used a more complex piece of equipment called a spectrophotometer to accurately assess how the flowers were absorbing and reflecting light in different parts of the colour spectrum. I then asked them to think about a question: are these flowers likely to be pollinated by insects or birds?

I was able to pose that question because the horticultural legacy of Britain means that some of the plants that we grow in our gardens are, in their native range at least, pollinated by birds. In cultivation, the flowers maintain that evolutionary history unless they have been highly bred and selected for by plant breeders, such that they hardly resemble their ancestors. A good example of this is seen in the fuchsias. In their wild condition, plant species such as *Fuchsia magellanica* produce elegant pendulous blooms, vivid red and sombre purple, that attract hummingbirds in their native South America. We still see these in gardens and in areas where the plant is naturalised, such as the west of Ireland, where ecologists Dara Stanley and Emmeline Cosnett have documented bumblebees stealing nectar.[1] The bees do this in two ways: by making holes at the base of the flowers into which they can insert their tongues, or by crawling past the dangling sexual parts of the flower and entering from the front. The latter technique may occasionally result in pollination occurring, but really the flowers have evolved to be effectively serviced by hovering hummers rather than crawling bumbles. Selective horticultural breeding and hybridisation, however, has turned out cultivars of this fuchsia in a range of colours, including white and pale lavender, and with bloated 'double' flowers. They're a far cry from the elegant ancestral species form, which luckily was widely grown on our campus in Northampton.

After analysis and discussion with the students, it was clear that there were significant differences between the two groups of flowers. Those that are (or were) bird-pollinated, such as the fuchsias, kniphofias from South Africa and New Zealand flax, tended to have larger, more robust, and more brightly coloured flowers, and to be unscented. Not only that, but the nectar was usually much more abundant and contained less sugar, compared with insect-pollinated lavenders, legumes and so forth. It was good to be able to demonstrate to my students that the imprint of a species' evolution may remain fixed even as we have changed the context in which that species is observed.

Those traits – structurally strong, of reddish colouration, unscented, and with copious, rather dilute nectar – characterise the classical notion of what bird-pollinated flowers are expected to be like. These expectations, in turn, emerge from what are termed the 'pollination syndromes'.

1 It was a struggle, because I do love a good pun, but I resisted the urge to call this chapter 'In the eye of the bee holder'.

Syndrome thinking

A pollination syndrome is a set of flower characteristics that, in theory at least, predicts the pollinator(s) of a particular flower.[2] Sometimes it works and sometimes it doesn't, but how often the shape, colour and scent of a flower leads to a successful prediction was, for a long time, hardly tested. Indeed, the pollination syndromes were accepted rather uncritically by many scientists and naturalists as unerring statements of fact. Though the original descriptions of the syndromes were drawn up in the nineteenth century, it was in the 1950s and 1960s that the syndromes were established as a way of, in effect, categorising plants using a 'taxonomy' that reflected convergent evolution, rather than traditional systematics of the type I described in Chapter 2.

Here's the thing about the syndromes: they can predict a flower's pollinators, but only for some species. Take, for example, the Argentinian cactus *Echinopsis leucantha* that I mentioned in the previous chapter. The species name *leucantha* means 'white-flowered' and, as you can see in Plate 8, it's an apt description. In the absence of any information about its pollinators, a cursory assessment of its white, tubular flowers might predict that they are pollinated by moths. However, when Pablo Gorostiague and Pablo Ortega-Baes conducted a detailed study of the plant, including night-time surveys, they found limited evidence of moth pollination. Instead, it was Gray-hooded Sierra Finches that were the effective pollinators. It's possible that moths may play a role in some populations in some years, but they are certainly not the only animals on which this cactus depends.

There are plenty of other examples like this, and I could probably write a whole book about the history and limitations of the pollination syndromes. But describing one more will help to really demonstrate how it's possible to get things wrong. In South Africa there exists a group of about 15 species of unrelated plants, from

---

2 Remember the distinction between a pollination *system*, which considers flower characteristics + pollinators, and a pollination *syndrome*, which is concerned only with the characteristics of the flower. For further discussion, see page 16 of *Pollinators & Pollination*.

different families, which have red, tubular flowers that produce large amounts of dilute nectar. These, as I noted above, are all characteristics of bird pollination. But when my friend and colleague Steve Johnson studied them in the 1990s he discovered that they are actually all pollinated by a single species of butterfly. Again, adherence to a strictly syndrome-based view of the flowers would not have recognised this without the hours of field observations that are required to test the predictions. We don't actually know how often bird-pollinated plants fit our expectations of what the flowers should look like according to traditional syndromes, because most plants have never been studied in that way. Often we're reliant on hypothesising based on the flowers. But hypotheses are just that: they are not statements of fact.

For every example such as the two I've just presented we can find cases where the syndromes do work, and the scientific literature on this topic is full of claim and counter-claim, example and counter-example, such that there's an element of yes-but-what-about-ism to the whole discussion. Part of the problem is that it's possible to choose a group of plants that fit our expectations of particular syndromes and go out and find what we expect. What that doesn't tell us is how often we'd correctly predict the pollinators of a plant chosen at random.

In the early 2000s a few of us tried to look at this more rigorously. In Australia, Andrew Hingston and Peter McQuillan addressed the question 'Are pollination syndromes useful predictors of floral visitors in Tasmania?' Their overall conclusion was that they were 'unreliable'. Amongst other things, Andrew and Peter found that flower-visiting birds mostly exploited the nectar of pale-coloured flowers, in contrast to the expectations of pollination syndromes – which would predict that they'd mainly go to more vividly coloured species. Andrew later added his expertise to a study I led, which we rather grandly titled 'A global test of the pollination syndrome hypothesis'. In this, we focused on six plant communities on three continents in which we had assessed the pollination syndromes of all plants in flower, according to the standard criteria of the 'classical' syndromes. To cut a very long story short, we found that on average the syndromes correctly predicted the flower visitors of about 30% of the plants. However, this varied enormously depending upon the syndrome and the situation. For example, in Peru bird-pollinated flowers were successfully predicted about 60% of the time, whereas in South Africa this dropped to about 8%. It's a study that is not without its flaws, but nonetheless it generated a lot of interest in the question of the effectiveness of pollination syndromes as predictors of a plant's pollinators. That was the point of the work: to get other researchers to start questioning and challenging ideas that had been accepted for decades.

Since then, the whole topic of pollination syndromes, how we should study them, and their practical utility, has become something of a minefield. Scientific disagreements are fine, and there are aspects of this whole topic on which my views differ from those of close colleagues. Sometimes, however, the debate has resulted in more heat than light, as when I was accused of lying at a Royal Society meeting by a very angry senior figure in the field who had misheard my response to a question that he'd posed.[3] If you're interested in investigating the subject of pollination syndromes more deeply, I've listed some key publications in the sources and further reading for this chapter. But I warn you, once you start reading this stuff, it's a bit of a rabbit hole.

Does it matter?

At this point we might ask the question: why does it matter? Is this really a subject important enough to get hot under the collar about? There are several reasons why it matters to me and why I'd prefer to see the pollination syndromes used more cautiously rather than as a defining framework for understanding the relationships between flowers and their pollinators. For one thing, the terminology is inaccurate, confusing and exclusionary. We traditionally add the suffix '–phily', from the Greek for love, to the names of these syndromes. So when we talk about ornithophily,

---

3 Fortunately the meeting was recorded, and afterwards I was able to reassure myself that I had not misled the audience. But it's a good example of how challenging accepted ideas (dogmas?) in science can land you in hot water.

we're literally referring to flowers that are 'bird loving', as though there was a deep and somehow meaningful connection between the two species. Not many people, even scientists working in the field, seem to appreciate that an aspect of teleological thinking was implicit in the development of the syndromes – in which some kind of 'grand pattern' was assumed to underlie the relationships between flowers and pollinators, rather than any such relationship being a result of strictly Darwinian evolution.[4] One of the results of this was a tendency for biologists to dismiss the importance of animal visitors that the syndrome didn't predict, or to describe them as 'secondary pollinators'. But many flowers that appear to be adapted to bird pollination may often also be pollinated by other animals such as bees. If we ignore these or relegate their importance, we're missing out on some really important ecology. It also provides us with only a partial, and rather oversimplified, understanding of how flowers evolve.

Principles revisited

In 1970, the renowned American evolutionary biologist George Ledyard Stebbins published a review of how flowering-plant reproduction evolves, and he included the statement that a flower's colour, shape, scent and so forth 'will be molded by those pollinators that visit it most frequently and effectively'. This has come to be known as Stebbins' most effective pollinator principle (or MEPP for short), and it seems to make sense: of course flowers will adapt to those pollinators that move the most pollen and ensure their reproduction, right? Well, yes and no. The MEPP certainly works for many species, but it's not the only principle in town. In the early 2000s Paul Aigner demonstrated, both theoretically and with data from studies of real plants, that flowers could become adapted to a *less* effective pollinator. That appears counterintuitive, but actually the logic is sound. Evolving, say, a flower colour or shape that allows a second type of pollinator to service the flower could be

4 Teleology, for the uninitiated, is a branch of philosophy that seeks to explain phenomena in terms of their purpose, as opposed to what caused them.

advantageous if at the same time this novel adaptation does not interfere with the effectiveness of the original pollinator. Let's refer to this as Aigner's least effective pollinator principle (LEPP).

So which is the principal principle, the MEPP or the LEPP? Which of these is more common in nature? Actually, we don't know because there have been too few rigorous, detailed studies of flowers that have sought to test between the two. However, there's a growing body of evidence that points to the LEPP being more common than we realise. Consider, for example, that survey of the diversity of pollination systems in the dogbane and milkweed family that I referred to in Chapter 4. This study revealed that about 7% of the species for which we had data possessed 'bimodal' pollination systems in which there were two distinct types of pollinators, such as bee + butterfly or butterfly + hawkmoth. These sorts of bimodal pollination strategies are not unique to that family. Decades of study by Nick Waser, Mary Price and colleagues at the Rocky Mountain Biological Laboratory (RMBL) in western Colorado confirm that Scarlet Gilia, a member of the phlox family with red tubular flowers that are almost archetypically 'ornithophilous', most often attracts hummingbirds. But in occasional years and locations, long-tongued bumblebees and other insects also visit the flowers, sometimes frequently. Working with Nick and Mary, Margie Mayfield found that flower stigmas are, on average, loaded with three times as much pollen after a single bumblebee visit as after a single hummingbird visit, and produce four times as many seeds. In other words, bees are better pollinators per visit but birds visit more often overall; the quality and quantity components of MEPP do not align. This kind of detailed experimental study requires many hours of patient observations and has been attempted in only a handful of plants, so we don't know how often the LEPP might be driving flower evolution compared to the MEPP. However, we do know that birds can pollinate flowers that are not obviously adapted to them – see for example the study conducted by some of my Brazilian colleagues, the title of which neatly summarised their findings: 'Hummingbird contribution to plant reproduction in the rupestrian grasslands is not defined by pollination syndrome'.

What is and what should never be

In 1969 British rock band Led Zeppelin released their second album. One of my favourite tracks from that record is called 'What is and what should never be'. Although the song itself is (allegedly) about an illicit romance, the title has always intrigued me as a description of both the possibilities of evolution and the ways in which we challenge scientific dogma. Natural selection has a way of coming up with solutions to problems or situations that allow species to exploit previously untapped opportunities. Whales lost their legs and evolved fins because it allowed them to be more effective denizens of the world's oceans. Likewise, birds evolved sophisticated anatomies that enabled flight. However 'what is and what should never be' might also be a description of how scientists sometimes constrain their thinking about these evolutionary possibilities. If we restrict our view of flower evolution and pollination biology to the expectations of the pollination syndromes, we risk short-changing nature.

A good example of this can be seen in the acacias, members of the pea and bean family (Fabaceae) that are often dominant trees in subtropical and tropical savannas. Most acacias are insect-pollinated but there are a few Australian species that utilise birds such as honeyeaters, thornbills and Silvereyes. In the Sunshine Wattle and the Golden Wattle, the flowers are whiteish or yellow and do not produce nectar. Instead, the birds feed from the sugary secretions of bright red glands that are located at the base of the leaves. These 'extrafloral nectaries' in other acacias have a role in rewarding ants that defend the trees from the depredations of herbivores. Such adaptations, in which the vegetative parts of a plant play an important role in pollination, are barely considered by the classical syndromes, which have traditionally focused largely on the floral structures.

For more than 50 years, the pollination syndromes provided a useful framework for understanding the ecology and evolution of plant–pollinator relationships, but in the last couple of decades we've learned that they are far from infallible as descriptors of patterns in the natural world. We can often say that a particular flower *is* bird-pollinated but cannot say that another is *not*. One of the challenges of assessing bird pollination in nature is that flowers may be very infrequently visited. Using a combination of direct observation and camera traps to study sunbird pollination in an endangered South African orchid, Timo van der Niet and colleagues concluded that they might expect to see one interaction every 13 days, on average, though in five of the nine years in which they tried to record these interactions they saw no flower visits. Our study of the bird-pollinated flowers on Tenerife – which opened Chapter 2, and which I discuss in more depth in Chapter 12 – came to similar conclusions. Considerable investment of time is required to properly document and study the pollinators of flowers, so syndromes provide us with a useful shorthand and a working hypothesis. But it's fundamentally important to test these assumptions and use the syndromes to generate hypotheses with testable predictions, rather than using them as statements of how the natural world evolves and functions ecologically. With this in mind, let's shift focus slightly and consider how the relationship between pollinating birds and their flowers functions as an exchange of goods and services.

CHAPTER 6

Goods and services

Whenever I visit a botanic garden I make a point of looking out for the flowers of a particular plant, in fact the one which adorns the cover of this book: the Bird of Paradise. This distant relative of gingers and bananas, with leathery, rather unkempt leaves, produces the most amazing bird-pollinated flowers. They are not only vividly coloured, but hide a surprising mechanism for ensuring pollination. I'm fond of demonstrating it to people who might not be aware of how these flowers function. First I show them the copious nectar spilling from the centre of the flower, below the orange petals, and get them to taste its sweet promise of a reward for services rendered. I also explain where the stigma is located, a sharp point standing out from the bloom, waiting for pollen to arrive. Then I ask them to locate the male parts of the flowers. There's often some puzzlement, because their location is not obvious. After some discussion I reveal the secret by asking them to push down on the blue petals, which open lengthwise to reveal the sticky pollen being dispensed by the anthers. This, I point out, is clearly a flower that has evolved a perch for birds that allows them to access the nectar. At the same time, this structure places pollen on the feet or belly of the perching bird.

The Bird of Paradise employs a sophisticated mechanism which, in its native Southern Africa, ensures seed set via the services of its main pollinators, perching birds such as the Cape Weaver. In North America, where the plant features regularly in landscaped gardens, hovering hummingbirds are poor pollinators of this plant. They will often take nectar from the flowers without picking up or transferring pollen, and when they do perch they may not be heavy enough to expose the pollen. However, perching New World warblers like the Common Yellowthroat are effective.

This interplay of flowers providing a reward for pollination is an example of what we might term 'biological barter', the trading of goods supplied (nectar, or sometimes pollen, that feeds the birds) for services rendered (ensuring the reproduction of the plant). In this chapter I want to explore this barter from two sides: how the birds move pollen around, and the rewards that they receive for their service.

Feathers and feet, tongues and beaks

Seeds are in many respects the ultimate expression of plant sexual reproduction, and for most of us this is what we consider to be the output of pollination. It's easy to forget, though, that flowers are usually hermaphroditic,[1] and can reproduce (in the sense of supplying the genes) as either 'mothers' or 'fathers'. It's the ovules of a flower, protected in an ovary and linked to the outside world via the pollen-receptive stigma, that provide that female function. That set of female structures can, in effect, make decisions about which pollen it will allow to fertilise an ovule, based on the origin of the pollen: for example, has it come from the same plant, does it have poor-quality genes, is it from a different population or species?

When it comes to the male function of flowers, things are arguably even more complicated. As well as serving a reproductive function, pollen is used by some flowers as a reward for those pollinators that can digest its protein-rich contents. The external wall of a pollen grain is largely indigestible (and very persistent in the environment, which is why fossil pollen is so useful for understanding

1 A small proportion of plant species are termed 'dioecious' and have separate male and female individuals within the same population. There are also 'monoecious' plants, which have separate male and female flowers on the same individual, in which case that *individual* is a hermaphrodite even if the flowers aren't. Then there are some very rare circumstances in which plants are 'andromonoecious' (both male and hermaphrodite flowers on the same individual) or 'gynodioecious' (some individuals with just female flowers, others hermaphroditic) or 'androdioecious' … You get the idea: plant sex can be very, very complex.

how plant communities have changed over time). However, many flower visitors have evolved ways to break into pollen and digest the contents, and a strategy used by pollen-rewarding plants is to produce far more than could ever father seeds – an approach in which the plant floods the market in the hope that at least some of the pollen serves its sexual rather than nutritional function. This kind of pollen rewarding is rare in bird-pollinated plants, as we'll see later, but when it does occur the result is often that pollen ends up all over the bird's face, head and other parts of its body.[2]

At the other end of the paternal spectrum are those flowers that place their pollen, delicately and precisely, on just one part of the body of their pollinating bird, often in a position where the animal cannot groom it off. This is observed in some tubular, horizontally orientated hummingbird-pollinated flowers where the anthers are positioned at the top or bottom of the flower. As the bird feeds, the anthers deposit pollen on their heads, under their chins, or on their backs. As German researchers Petra Wester and Regine Classen-Bockhoff have shown, in some members of the sage genus *Salvia*, a lever mechanism means that only those birds of an appropriate size and behaviour are dosed with pollen.

The ideal pollinator, as far as the flower is concerned, is one to which pollen can stick persistently, but not *too* persistently: it must be able to rub off when a suitable stigma of another flower is encountered. That 'stickiness' is determined by a number of different factors, including chemical coatings on the pollen or its electrostatic charge, which bind it temporarily to the bird. If you explore Bird of Paradise flowers as I described at the start of the chapter, and pick up some pollen on your fingers, you'll see that it's entangled in very fine material that we call 'viscin threads'. These threads serve to attach the pollen to the birds' undersides, an unusual part of the body for a flower to exploit. The exact point of pollen placement is crucial: the rougher or more feathery, the better. In bird-pollinated flowers, pollen often ends up on the feathers of the avian visitor, where it is captured by the complex of barbs and barbules.

---

2 There are other possible explanations for this, however, including male–male competition, which are beyond the scope of this book.

Although feathers would seem to be the obvious evolutionary choice for where to place pollen, in fact there are many examples of alternative flower strategies in which beaks, feet or even tongues are involved. There are orchids, for example, that have a foot fetish, and which Steve Johnson first described in a South African species almost 20 years ago. Orchids are unusual in that they bundle their pollen into more or less cohesive packages attached to adhesive structures. But they're not the only plants that do this.

When pollen comes packaged

When we think of pollen, what usually comes to mind is a yellow or orange powder that may set off allergies (in the case of those who are sensitive to the pollen of wind-pollinated plants) or which can stain clothes if you misposition lily flowers. But pollen comes in many forms, depending on the plant: it can be virtually any colour from white to the darkest of purples; it may be dry and dusty or wet and sticky; and it varies hugely in shape, and in size from microscopic to individual grains that can be seen without the aid of magnification. Also, in some rare cases, the pollen is packed into dispersal units called 'pollinia' (singular = pollinium) that contain hundreds to thousands of individual grains. I say 'rare' because pollinia have evolved only a handful of times, but in fact they are the hallmarks of two rather large plant families. The first is the dogbanes and milkweeds (Apocynaceae) that we met in Chapter 4, which, with around 5,350 species, is about the tenth largest family. The second is the orchids (Orchidaceae), which, with about 28,000 species, vies with the daisies (Asteraceae) as the largest plant family.

These two pollinia-bearing families are only very distantly related, and this type of pollen packaging has thus evolved independently in these lineages. This is a clear example of convergent evolution, where both plant groups have come up with the same solution to a problem. In this case, the problem may be that bees collect a lot of pollen, and this has a reproductive cost for plants. Packaging the pollen in this way prevents bees from stealing it, amongst other advantages.

The pollinia themselves are very different too, and we could get into some very detailed botany trying to explain their morphology and functioning, so what I'm about to tell you is the simplified version. In orchids the pollinia are connected to a sticky 'viscidium' that attaches them adhesively to a pollinator. If you grow moth orchids (*Phalaenopsis*) as a house plant you can find the pollinia in the centre of the flower, underneath a central column, which is the fused male and female parts of the flower.

In the dogbanes and milkweeds, it's the latter group that have the pollinia, and we refer to them as 'asclepiads' because they are members of the subfamily Asclepiadoideae.[3] Asclepiad pollinia clip on to a pollinator using a mechanical structure that links two pollinia together.

Whether it's an orchid with its sticky viscidium or an asclepiad with its clip mechanism, we refer to the whole (pollinium + mode of attachment) collectively as a 'pollinarium' (plural = pollinaria).[4]

Observations of orchids from the nineteenth and early twentieth centuries recorded birds as visitors to their flowers. However, it was not until the 1990s that it was shown categorically that birds can be effective pollinators of some species (Plate 26). According to recent reviews of their pollinators, at least 3% of orchids (about 1,000 species) are confirmed or suspected to use hummingbirds to move their pollinaria, principally glued to their smooth beaks. Intriguingly, these pollen packets tend to be more sombre in colour (brown, grey or blue) rather than the usual bright yellow, which is thought to make them less conspicuous to the birds and therefore less likely to be groomed off. In addition to hummingbirds, we can add sunbirds to the list of orchid pollinators in southern Africa, and also white-eyes on the island of Réunion in the Indian Ocean.

I suspect, however, that bird pollination in orchids is under-documented, for a few reasons. First, it's rarely looked for because

3 The milkweeds (Asclepiadoideae) used to be a family in their own right, the Asclepiadaceae. But that's another botanical rabbit hole that we won't venture down.

4 There are actually a few Apocynaceae that have evolved sticky pollinaria structures with which to move their pollen, but for now we'll not overcomplicate an already convoluted story.

it seems improbable. What plant is going to glue a pollen mass to a bird? How effective is that going to be as a mode of reproduction, especially as many orchids are naturally rare and have low population densities? But secondly, and perhaps more fundamentally, it's rarely observed because the observer needs to be in the right place at the right time to see it. Bird pollination of orchids can be very infrequent, and as I noted in the previous chapter, even with modern camera traps, the chances of documenting bird visits to orchids can be low. Although aspects of flower colour, size and shape can suggest bird pollination in some cases, it doesn't always work. Those white-eyes on Réunion, for example, pollinate orchids with white flowers that we might expect to attract moths rather than birds. Indeed, as botanist Claire Micheneau and colleagues have documented, other members of this particular group of orchids are moth-pollinated, so the white flowers seem to be an evolutionary hangover. It's a good example of what I described previously, that whilst we can often successfully predict that a flower *is* bird pollinated, it's much harder to predict that it's *not*.

What about those mechanically clipping asclepiads? Are they ever bird-pollinated? If you'd asked me that 30 years ago, when I first became interested in the ecology and evolution of pollination systems in this group of plants, I would have said that it was unlikely. Pollination of asclepiads usually requires a fairly precise orientation of the pollinaria on a pollinator, in relation to the morphology of the flower. I would have predicted that attaching pollinaria to feathers, which by their nature are flexible and rather mobile, would be too imprecise. But in 1998 the journal *Nature* asked me to review a paper by pollination ecologist Anton Pauw that forced me to revise that view. In this study, Anton showed that a South African asclepiad is indeed pollinated by the Southern (or Lesser) Double-collared Sunbird. But the pollinaria clip not onto the bird's feathers, but to its tongue! They're able to do that because the tips of the tongue are finely frayed (to mop up nectar, of course) and narrow enough to act as points of attachment for the pollinaria. Based on this, Anton predicted that hummingbirds and honeyeaters might also pollinate asclepiads in other parts of the world. That has never been tested, to my knowledge, though hummers certainly visit milkweeds in North America.

After Anton's paper was reviewed and accepted, *Nature* asked me to write a commentary about it in their 'News and views' section, which I was happy to do. I called it 'Sunbird surprise for syndromes' and used Anton's ground-breaking study as an example of the wider issue of predicting pollinators from assessments of flower characteristics. Although some features of the flowers of this sunbird-pollinated asclepiad fit the bill of bird pollination (if you'll excuse the pun), for example the reddish petals and the absence of scent, other aspects are less characteristic. And of course the fact that these flowers present their pollen as discrete pollinaria is the biggest surprise of all. The nectar that's produced by this asclepiad is fairly large in volume and continuously secreted, and on the dilute side, though not excessively so. That's a good point at which to turn to the question of nectar and why birds are so keen to accept it as a reward.

Why nectar?

When you use a sugar refractometer to measure nectar concentration in the field, you have to act fast, especially if the weather is hot and dry. If you don't quickly search for the point at which the line crosses the scale, the evaporation of water from the glass prism can alter the reading. Then you need to clean it carefully before putting on the next sample. My usual technique is to justify my hard work with a sweet reward, and lick the nectar from the glass, before cleaning and drying it with a cotton handkerchief. Usually that works well, and I've done it hundreds of times. The one occasion that it's been a problem was on Tenerife when I was collecting data for our study of bird-pollinated plants on that island (see Chapter 12). The Canary Island Foxglove, it turns out, has nectar that tastes slightly bitter, and it didn't take long for my tongue to start tingling and my mouth to feel numb, like the after-effects of a visit to the dentist. This plant apparently loads its nectar with chemical compounds that, to humans at least, are repellent or even toxic. That might explain why we rarely saw bees probing the flowers. However, it was a nice demonstration that nectar is far more than just sugar dissolved in water.

At the end of Chapter 4 I asked the question, why birds? What's the advantage to flowers in being bird-pollinated? I can just as easily turn that question around and ask, from the birds' perspective, why nectar? What's the advantage of incorporating some, or a lot, of this sugary fluid in their diet? First and foremost, of course, it's about energy: birds are warm-blooded, often fast-flying, highly active animals. They need a lot of energy, and nectar can provide a very quick boost, much like a glass of lemonade after a period of hard graft in the garden. Nectar contains more than carbohydrates, however, and it can't be emphasised enough just how complex it is both as a substance and as a tool in a plant's strategy for reproduction. The main constituent of nectar is of course water. But water itself is an important resource for birds and may be especially vital in the dry seasons of warmer parts of the world, when some bird-pollinated plants flower most frequently (see Chapter 8). Without water, these birds cannot metabolise any food that they consume and therefore cannot make it through the dry season.

In addition to water, the next most abundant chemicals in nectar are energy-rich sugars, of two main types: hexose sugars, with six carbon atoms, such as glucose and fructose, which are referred to as monosaccharide sugars; and sucrose, a disaccharide sugar formed from a combination of glucose and fructose molecules. Sucrose is the sugar with which most of us are more familiar, usually refined from crops such as sugar beet or sugar cane. Depending on the relative proportions of these sugars, nectar may be described as being 'hexose-dominated', 'hexose-rich', 'sucrose-rich' or 'sucrose-dominated', with the ratio of sucrose to hexose varying accordingly. This is really sucrose/(glucose + fructose) but let's refer to it as the S:H ratio for simplicity.

In 1990, husband-and-wife biologists Herbert and Irene Baker suggested that S:H ratios were predictive of different types of pollinators. Their work showed that, for example, hummingbird-pollinated species of coral tree in the pea and bean family

(Fabaceae) usually had sucrose-dominated nectar (average S:H = 1.30), whereas passerine-pollinated species (from both the New World and the Old World) produced hexose-dominated nectar, with average S:H of 0.04. Similar results were found for members of the pineapple family (Bromeliaceae) that were also either hummingbird- or passerine-pollinated.

In addition to generalisations about sugar ratios, research by the Bakers and by others suggested that there were additional ways in which hummingbird flowers differed from flowers pollinated by other types of birds. Specifically, that natural selection by hummingbirds had resulted in flowers that tended to produce relatively large amounts of rather dilute nectar. A bee-pollinated flower, for example, might contain a standing crop of nectar (the quantity of nectar available in a flower at any one time) of less than 1 microlitre, with a sugar concentration in excess of 30%. Bird-pollinated species can have at least 10 times that amount of nectar, and less than 20% sugar is more typical. These figures for volume and concentration are broad generalisations, however, and in fact such values can be hugely variable within a plant population, and even over the course of a day for a single plant.

The orchids often buck this trend by producing only very small quantities of nectar or sometimes none at all, in which case they rely on naive birds that have not yet learned to avoid the unfulfilled promises of those flowers.

In the decades since the Baker & Baker paper and other ground-breaking work, pollination ecologists have discovered that things are a little more complicated than the textbook examples would suggest. This is perhaps not surprising, given that only a small proportion (perhaps 10%) of the flowering plants have had their pollination systems investigated. So it was that in 2008 South African ecologists Steve Johnson and Susan Nicolson challenged the received wisdom when they wrote:

> A longstanding paradigm in biology has been that hummingbirds and passerine birds select for different nectar properties in the plants they pollinate. Here we show that this dichotomy

> is false, and a more useful distinction is that between specialized and generalized bird pollination systems.

What Steve and Susan discovered was that South African plants that are specialised for sunbird pollination, such as some aloes, also have sucrose-dominated nectar, and in similar high volumes and low concentrations. It's a good example of how long-held assumptions about plant–pollinator interactions, indeed about ecology generally, can be overturned by more detailed studies and rigorous tests of those ideas. In this case, however, it also raises the question of what we mean by 'specialised' and 'generalised' bird pollination systems. For some researchers this refers to the degree of dependency that a bird has on the nectar provided by flowers. However, reliance on nectar spans a broad spectrum. Hummingbirds are completely dependent on nectar to provide energy, but not for essential amino acids and minerals, so all species include insects, spiders and other arthropods in their diet. At the other end of the scale, the tits and warblers that are occasional flower visitors can almost certainly survive without access to flowers. But between these extremes are whole groups of birds whose dependency has never really been assessed. For example, we believe that all of the white-eyes, yuhinas and their relatives (Zosteropidae), more than 140 species in total, visit flowers and feed on nectar. But how reliant are they on that nectar? If it's only important at certain times of the year, for example during the dry season when little other food is available, then I think there's a case to be made for describing these birds as specialised nectarivores, even if they never figure as such in the textbooks. Similar arguments can be made for other passerine families.

Another thing to consider is that even those highly specialised hummingbirds are not completely selective about the chemical constituents of the nectar on which they feed, and several studies have shown that they will forage from flowers that are hexose-dominated as well as sucrose-dominated, and often take nectar from insect-pollinated flowers that have much higher concentrations than are typically found in classically 'ornithophilous' flowers.

During a career spanning almost 50 years, Australian biologist Graham Pyke has thought about the complexities and paradoxes of nectar in bird-pollinated plants more than most. In 2019 I had the pleasure of spending time in the field with him, including a day of birding in the Royal National Park, south of Sydney. We discussed his perspectives on nectar, including questions of why concentrations tend to be low when experimental work with birds shows that (not unexpectedly) they prefer artificial 'nectars' with larger amounts of sugar. Indeed, Graham wrote recently that flower characteristics

> involving floral nectar have eluded satisfactory evolutionary explanation … perplexing pollination biologists and arousing recent controversy.

In addition to unanswered questions about the type and concentration of sugar in nectars, there are others regarding the role of amino acids and secondary compounds such as alkaloids, and how these can be altered by yeasts and bacteria, as we'll see in Chapter 10. Then there are enigmatic phenomena such as the rare occurrence of flowers with coloured nectar, which can be any shade from blue to red. Nectar can even be black. Recent research by Evin Magner and colleagues showed that the black nectar found in the flowers of a South African plant (Plate 9) is actually more conspicuous to pollinating birds than the red flowers. They also discovered that the nectar colour is the result of chemical reactions similar to those used by medieval monks to produce black ink from oak galls mixed with iron. So much for nectar being a supposedly simple substance.

Other rewards for a job well done

Pollen is a problematic reward for flowers to produce, because the primary function of pollen is of course reproduction and the passing of a male set of genes on to the next generation. Any pollen grain consumed is a lost opportunity for reproduction. Additionally, in comparison to nectar, pollen is relatively expensive for plants as it requires more energy, in the form of carbohydrates and lipids, and resources such as nitrogen, phosphorous and potassium. Pollen can therefore be a rich source of nutrients for bee larvae, *Heliconius* butterflies, syrphid flies and some other insects, and there are many groups of insect-pollinated plants that primarily offer pollen as a reward, for example the poppies. Some bat-pollinated plants also feed their furry flower visitors with pollen, often in addition to nectar.

What about the feathery flower visitors? As far as I am aware, there are no flowers that reward their bird pollinators only with pollen. Experiments with adult psittacines, hummingbirds and honeyeaters have shown mixed results. Honeyeaters can digest significant amounts of pollen, but it depends on the plant species, and most can pass through the gut undamaged. David Paton has argued that New Holland Honeyeaters ingest so little pollen that even if it was all digested it would make a negligible contribution to their diets. However, it may be more important to the diet of nestling lorikeets in Australia. Likewise, in the Galápagos Islands, young Common Cactus-Finches and Medium Ground-Finches are fed prickly pear cactus pollen by their parents in the dry season when there's a shortage of other food. According to Rosemary Grant, who has worked on these birds for decades, more than 90% of the pollen is digested. The little work that's been done so far on hummingbirds indicates that they cannot digest pollen at all.

The botanical dilemma of feeding a pollinator with a reward that could otherwise increase a plant's reproductive success has found a satisfactory solution in a few groups that have evolved flowers which produce two sorts of anthers. These contain different types of pollen that have divided the reward and reproduction functions between them. Some bird-pollinated plants go further and have evolved 'stamen bodies' in which some of the male reproductive organs have transformed into nutritious rewards for the birds. Agnes Dellinger and colleagues studied this in the South American genus *Axinaea* and showed that the large, bulbous modified stamens are actively consumed by tanagers and flower-piercers. Agnes and her co-authors beautifully describe how these flowers function:

> Using their beaks, the perching birds seized a bulbous stamen appendage, ripped it from the flower, and consumed the appendage together with the anther and part of the filament. We observed that upon initially seizing and compressing the appendage, a pollen jet was ejected, coating the bird's beak, forehead, or neck.

Who says scientific writing has to be dull?

Beyond pollen and stamen bodies, there are other ways in which flowers can feed birds in return for their pollen-shifting services. Some plants produce nutritious bracts or petals that the birds pluck and consume. One of the most intriguing examples is a species of slipper flower (*Calceolaria*) that Alicia Sérsic and Andrea Cocucci documented in Tierra del Fuego at the southern tip of South America. The genus gets its vernacular and scientific names from the way that the lowermost lip of the flower is folded and shoe-shaped – *calceolus* is Latin for a small shoe or slipper. In the species that Alicia and Andrea studied, the fold has become 'massive and fleshy' in comparison with most other species. Birds remove and eat this structure, in the process picking up or depositing pollen. Surprisingly, the pollinators are Least Seedsnipes, members of the order Charadriiformes, which contains waders and shorebirds, as well as gulls and auks. This is the only known pollinating bird from that order, though as we'll see in Chapter 11, recent work by those scientists has revealed that other birds are also implicated in this pollination system.

Let's end this chapter where we started, with nectar. But not nectar as we know it, rather nectar that has been transformed into a 'sweet jelly', in the words of Marlies Sazima and her co-researchers. This is how a bushwillow (*Combretum*) from the Pantanal of Brazil presents its nectar to almost 30 species of birds belonging to eight different families, of which the effective pollinators seem to be various orioles, thrushes, and tanagers. Each flower produces a single jelly pellet about 6 millimetres in diameter, which the researchers describe as having a sweet taste – hardly surprising, as it contains 90% sugar. Clearly, I'm not the only biologist who enjoys sampling nectar.

CHAPTER 7

Misaligned interests

It's day four of a COVID infection. Feeling much better, but still low on energy, I sit outside our tent trying to wash down dry bread, slathered in peanut butter, with hot black coffee. In the low morning light the Kenyan savanna glows russet and green, while the bark of a nearby Yellow Fever Tree, a type of acacia, shines golden. Though I'm feeling much better than I have in the last few days, raising my binoculars to view the birds and the landscape is still an effort. Thoughts of illness are temporarily forgotten, however, when I spot a pair of Speckled Mousebirds feeding on the flower heads of the acacia. Their delicate beaks gently remove small clusters of petals, swallowing them with a few deft movements, leaving most of the inflorescence behind. In contrast, the Vervet Monkeys I watched earlier in the week ate by tearing off whole flower heads and cramming them into their constantly chewing mouths.

As the birds move carefully along the thin branches, their belly feathers and tails were almost certainly picking up pollen from the densely packed flowers. Could these birds be legitimate pollinators, with the acacia rewarding them with flower tissue rather than nectar? Is that why the unopened buds are a deep crimson, to act as a strong visual signal? What about the roles of the many other flower visitors to this tree – such as the honey bees and the butterflies – do they contribute more to pollination than the birds? So many unanswered questions about a common bird and a common tree in this part of Africa. I scribble down a few notes and return to the tent to see if my wife, also bedbound with this damn virus, needs anything.

What's my motivation?

That question has become a thespian cliché, of course, but it makes a good starting point for considering birds and plants as actors in an ecological play. No bird wants to pollinate a flower, in the sense of it being a beneficial behavioural choice, akin to choosing a mate or building a nest or raising young. When it comes to flowers, all that birds are interested in is finding food, though the observation that I noted in Chapter 3, that the Crested Berrypecker is known 'to pluck large ericaceous flowers and smear themselves with them', suggests that the relationships between birds and flowers may sometimes be more complex than we currently understand. As with my Kenya mousebird observations, this raises so many questions. Why would berrypeckers do this? Does it help to attract a mate? Does it prevent parasites such as feather lice? Even in such well-studied organisms as birds and flowering plants, there's much to be discovered. In reality we have an iceberg understanding of biodiversity and ecology: the knowledge that's on the surface is just a fraction of what's hidden from us.

Whatever the motivations that drive birds to interact with flowers, the transfer of pollen from one bloom to another is simply a side effect of that instinct for obtaining something of benefit to itself, typically sustenance. Clearly, it's an instinct that bird-pollinated flowers exploit for their own good, and they have evolved ways to win in this relationship. But they have done so in the face of, at best, indifference and, at worst, active destruction on the part of the birds. The interests of the flowers and the birds that exploit them are thus misaligned. This conflict of interest in turn has resulted in the evolution of bird and flower behaviours and morphologies in which each party adapts the interaction to its own advantage.[1]

Flower–pollinator interactions are frequently described as 'mutualisms' in which both species benefit – and this is often, though not always, the case. However, mutualistic relationships

1 Yes, plants have 'behaviour', though our animal brains may not always recognise it as such.

are not the same as cooperative ones: mutualisms are really about mutual exploitation, and as in all exploitative connections, things can get out of hand such that one partner gains a lot at the expense of the other partner. In bird–flower relationships, this ranges from plants that offer no reward to avian actors that destroy the whole bloom. Somewhere in the middle is that (literal) sweet spot where birds gain a sugary reward from plants which in turn have their reproduction assured. In order to determine whether these interactions are truly mutualistic, we need to closely observe the natural history as it happens, and preferably experimentally manipulate the interactions to understand what's occurring.

How can we test whether flower visitors are effective pollinators?

When we watch birds or insects visiting flowers, preferably with the aid of binoculars, we might make assumptions about whether or not they are effectively transferring pollen to the stigma. Those assumptions may be based on the size and behaviour of the visitors in relation to the morphology of the flower, or, if we've been able to look closely, on whether or not the visitor actually touches the stamens during some visits and then contacts a stigma during others. Our assumptions may be correct or they may not; the only way to know for sure is to set up experiments. These come in various forms, but often it's about observing what happens when a flower visitor encounters a bloom that's newly opened. We can then count the number of pollen grains on stigmas after single visits to flowers – what's known as an SVD or single-visit deposition experiment. Or go a stage further and bag the flower with fine netting after a pollinator visit. This excludes other potential pollinators, but allows the circulation of air and the further development of the flower into a fruit containing seeds.

In the high Andes of Peru, a shrubby member of the verbena family (Verbenaceae) called *Duranta mandonii* is commonly encountered in dry forest and areas of wet cloud forest that have been disturbed by farming or felling trees for firewood and construction. It typically grows to less than 5 metres in height and produces small, orange, fleshy fruits that are consumed by birds – which then spread the seeds throughout this mountainous ecosystem. Prior to that, the plant is covered in small white tubular flowers, each only a centimetre or so in length. When Stella Watts studied this species in Peru as part of her PhD research with me, she discovered that the flowers individually produce only modest amounts of nectar, typically about 3 microlitres in a 24-hour period. The nectar is moderately rich in energy, containing about 20% sugar on average. However, these flowers replenish their nectar when it's been drained, and large plants can have more than 500 flowers open on any one day. Potential pollinators that choose to exploit them can therefore obtain a lot of nectar in a small area over a short space of time. This is a critical point: any animal needs to balance the energy obtained from a food source with the energy required to find and exploit that food. If the latter is greater than the former, the animal will starve. Biologists have studied this for decades in various groups of animals, including nectar-seeking pollinators, and there's a big body of research on what has been termed 'optimal foraging theory'. This theory predicts how animals will behave if they have chosen a diet and a way of obtaining their food that is best for them, in terms of reproduction.[2]

After spending some time observing patches of *Duranta*, Stella and her field assistants (myself included, for a period) observed that the flowers were visited by a wide diversity of insects and birds. These included several types of butterflies and moths, a couple of species of native bumblebees, some non-social 'solitary' bees, introduced honey bees from local hives, hoverflies and other Diptera, two species of hummingbirds (Tyrian Metaltail and the Shining Sunbeam) and the Masked Flowerpiercer. The latter is a member of a genus that, as I noted earlier, have specialised bills that they use to pierce the base of a flower and thus gain access to the nectar without pollinating, though recently their role as pollinators of some plants has been confirmed. In fact *Duranta* seems to have

2 More controversially, the theory has also been applied to plants which 'forage' with their root system or by producing clonal offsets. As I said, plants have behaviour.

evolved a strategy to get around the problem of flower piercing by flowerpiercers: the green sepals which surround a blossom when it is in bud are thickened as an evolutionary strategy that functions to deter this kind of nectar robbery. Indeed, during the course of our study we found no *Duranta* flowers that had been pierced in this way: any animal, be it insect or bird, that wants to drink the nectar is required to enter the flowers legitimately, from the front. Even the flowerpiercers.

The small, white tubular flowers did not immediately strike us as being adapted for bird pollination, even though birds were the most frequent visitors by a long margin. The Tyrian Metaltail on average made almost 70% of the flower visits in any one observation period, whilst none of the other species or groups accounted for more than 10%. But were these birds effective pollinators? To answer that question, Stella set up an experiment that involved covering flowers with muslin, then removing it at set periods and allowing the birds or insects to visit. For a sample of these flowers she counted the amount of pollen removed from anthers, which is a measure of the potential 'male function' that I described in the previous chapter. Another set of flowers were allowed to develop into mature fruit, which were then collected and the number of seeds counted.

Although I've summarised these observations and experiments in a few lines, they involved almost 10 weeks of field work in Peru plus many days of pollen and seed counting back in the laboratory in the UK. This kind of research is time-consuming: the resulting data set is hard won. But once the results were in, we discovered that despite the Tyrian Metaltail hogging most of the nectar, due to its frequent flower visits, this bird removes only about 1% of the pollen from anthers. Not only that, but it contributes precisely *zero* to seed production. In contrast, we found that visits by native bees remove half of the pollen, and that these insects are highly effective at pollinating flowers and so initiating seed production.

The reason for the discrepancy between how often the birds visit the flowers and their lack of pollinating ability is fairly obvious: in these short-tubed flowers, pollen can only adhere to the tip of the smooth bill, from where it easily falls off, or is groomed off. If the anthers contacted the bird's head feathers, as those of the cactus in Plate 10 do, it would be a different story. Although bill-tip pollen transfer does work for some plants, it seems to be ineffective in *Duranta*.

I think it's important to emphasise again that it requires these sorts of detailed observations and time-consuming experiments to work out which pollinators are effective. It's very labour-intensive, and in this example it involved working and living in challenging conditions at elevations of between 3,200 and 3,850 metres above sea level in the Andes, even before Stella could start to dissect flowers and fruit back home. All of this before the data could be analysed and the results written up, first for Stella's thesis, then for publication in the journal *Plant Species Biology*, where it appeared in 2012, almost a decade after the first data were collected. Science is usually a marathon, not a sprint.

Angry birds!

Our conclusion from the *Duranta* study was that the flower-visiting birds were not pollinators: they were stealing nectar from the plant but not transferring pollen. But does that mean that they were playing no role in the reproduction of *Duranta*? Not directly, but we know from studies of other plants that there are indirect ways in which nectar thieves can positively affect plant reproduction. One is that, by reducing the amount of nectar available in each flower, the nectar thieves force the legitimate pollinators to probe more deeply, increasing the chances that pollen will be deposited on a stigma. Not only that, but nectar robbing can cause pollinators to move more often between different individual plants, thus distributing pollen more widely, as Vineet Kumar Singh and colleagues showed in the case of a self-incompatible bird-pollinated tree in India.[3] How often these mechanisms occur in nature has not yet been determined.

There is one other way in which nectar-robbing birds can positively influence plant reproduction, though again we don't know how common it is. By aggressively driving off the real

3 'Self-incompatible' means that the flowers cannot be pollinated by pollen from the same individual plant.

pollinators, nectar thieves can potentially increase the movement of pollen between flowers, spreading a plant's genes further within and between populations. I first observed this behaviour in African sunbirds during the easiest field work I ever conducted. Comfortably seated on a wooden veranda, overlooking a patch of beautiful rainforest, the table behind me loaded with tea, coffee and snacks … data collection does not get any more luxurious! This was during the first time I taught on a Tropical Biology Association field course, at Amani in the East Usambara Mountains of Tanzania. I spent time watching the behaviour of a male Eastern Olive Sunbird as it defended a patch of flowers from any and all incomers. It carefully drank from the open flowers but managed to miss the sexual parts. Female birds were tolerated, in exchange for mating rights, but other males were quickly seen off. The bird particularly disliked the large carpenter bees that made an occasional entrance. Once a bee was spotted taking nectar from a flower, the sunbird flew at it and chased it away into the forest. Long after they were out of view I could hear the click-click of the sunbird's beak as it snapped against the bee's hard exoskeleton.

That sunbird was robbing nectar from a large scrambling vine, a wild member of the squash and melon family, and based on their behaviour we can be fairly sure that the bees were the actual pollinators. By driving the bees away from the flowers just after they had picked up pollen, the birds were potentially helping with the reproduction of the plant. The bees would be more likely to visit the female flowers of this dioecious species, which incidentally closely resemble the male flowers in what is probably a case of within-species mimicry, with females pretending to be males. The female flowers do not, however, provide any nectar and so are not defended by male sunbirds.

In the case of the sunbird–vine–carpenter bee interactions, the implications for all of the species involved were considerable, affecting both sources of food and opportunities for reproduction. These three species formed a simple network that was embedded within a wider, complex, dynamic set of relationships between flowers and insects and birds that extended beyond the field centre where we were based. I'll come back to this idea of networks and other players in the plant–pollinator relationship later in the book, but before we do, let's return to flower eating.

Florivory in all its gory detail

Flor-I-vor-EE or *flor-IV-or-ee* is the consumption of flowers. Many animals, including birds, rodents, primates (including humans), snails and insects, eat flowers. It's a surprisingly common way of obtaining a meal, even if we don't always notice it. On a recent trip to southern Spain, sitting beneath a canopy of café parasols, I lost track of the conversation around me as my eyes were drawn to the movement of reddish, glossy leaves and garish pink flowers. Something was bustling in one of the potted begonias anchored in its bright ceramic pot to the whitewashed wall of a nearby building. A sparrow – but were they House or Spanish Sparrows? Another bird joined the first, and it was only then that I realised they were eating the begonia flowers. It was hot – 36 °C in the shade, an azure sky and barely a breeze – and these thirsty birds were making the most of a succulent floral snack. Such relationships, it seems, are everywhere if we take the time to observe.

The most comprehensive study of florivory to date, by Maria Gabriela Boaventura and colleagues, implicated birds (alongside mammals) as the main flower-eating culprits. As I described at the start of this chapter, the outcome of florivory can be more or less damaging to the reproduction of a plant. Consuming the whole flower or inflorescence, as the Vervet Monkeys did, reduces to zero the chances of that flower setting seeds. But losing a few flowers from a flower head, or just a few petals or stamens, can still allow a plant to fulfil its reproductive destiny. Much like the Black Knight in *Monty Python and the Holy Grail*, a flower can withstand a surprising amount of damage inflicted by a florivore before it ceases to function.[4] The critical thing for seed set is that the animal

4 The same is also true of seeds, as I discovered during my PhD research: experiments with partially eaten seeds of my study species, Bird's-foot-trefoil, showed that they can lose more than half of their tissue and still successfully germinate and grow into plants. This has also been found in other plants: nature can be surprisingly resilient.

does not damage the stigma, style or ovary, though if by eating most of a flower the offending animal gets smeared with pollen, then the flower could still parent a seed on another plant, via its male function.

As discussed in Chapter 1, there's a longstanding debate as to whether florivory or feeding on flower-visiting insects was the precursor to pollination by birds. At the moment I don't think we understand enough about the consequences of eating flowers from a plant's perspective to really weigh the 'florivory hypothesis' against the 'insectivory hypothesis'. But I've always been sceptical of simple, single-answer explanations of ecological and evolutionary patterns, and it's quite possible that both of these were important in the past for different groups of plants and birds at particular periods over the last few tens of millions of years.

Delicate, long probing bills are frequently found in some groups of flower feeders, such as hummingbirds, sunbird-asities, honeyeaters, woodpeckers and sunbirds, but by no means all. Shorter, sturdier beaks better suited for plucking and tearing are also common, especially in the white-eyes, finches and flowerpeckers. Some flowers have evolved to exploit this less refined dining behaviour, and there are times when flowers need birds to vigorously manipulate them, rather than gently probing, in order to be pollinated.

The explosive flowers of mistletoes

When we associate mistletoe with Christmas traditions and pagan rituals, we're normally considering only one species, the European Mistletoe, though in the New World the native American Mistletoe acts as a stand-in for its European cousin. To botanists, however, 'mistletoes' really refer to a plant life strategy in which one plant is hemiparasitic on another. Hemiparasites have green leaves or stems and can photosynthesise to gain energy, but they receive their water and mineral nutrients such as nitrogen, potassium and phosphorous by tapping into the vascular system of their host.

This exploitative lifestyle has evolved many times in different plant families. In most cases we don't notice it because the hemiparasite taps into the underground root system of the host, Yellow Rattle being a good example in the northern hemisphere. The nature of the interaction is only obvious when the exploiter grows on the trunk or branches of a tree or shrub, and that's what we mean by a mistletoe.

Mistletoes have evolved independently in several families, though all are in the sandalwood order (Santalales), suggesting that there's something about this group that predisposes them to the mistletoe habit. Mistletoes are extremely important in the ecosystems in which they occur, leading Australian ecologist David Watson to describe them as a 'keystone resource'. Although the term 'keystone species' is rather overused, sometimes inappropriately, in conservation, in this case it's justified.[5] Many mistletoes provide food for birds, in the form of nectar and fruit, at times of the year when other resources may not be available, especially in the dry season or during prolonged droughts.

Bird pollination mainly occurs in a family that is aptly named the 'showy mistletoes' (Loranthaceae), a group that is widely distributed and especially common in the tropics and the southern hemisphere. As the name suggests, many of the showy mistletoes have flowers that are bright red, yellow and orange. In some of these bird-pollinated species, the flowers do not open of their own accord: they require a curious bird to tear or prise open the buds. Some New Zealand mistletoes that are pollinated by the Tui and the New Zealand Bellbird (both honeyeaters) possess these 'explosive' flowers. Researchers Jenny Ladley, Dave Kelly and Alastair Robertson describe this dramatic flower strategy as follows, with different species possessing either 'one-stage' or 'two-stage' opening mechanisms:

> In two-stage opening, the foraging bird first
> grasps the bud in its bill and gently squeezes

5 Describing a species as a keystone implies that it has a big, positive effect on the other species with which it shares its habitat, despite the fact that it is in relatively low abundance. So oak trees in an oak forest and earthworms in a grassland are not keystone species, even though they benefit other species.

> or twists. This causes the petals to explode open into a symmetrical flower. The bird then inserts its beak into the open flower and takes the nectar … In one-stage opening, the foraging bird inserts its bill into one of the … splits between the petals … of a mature bud. This causes the petals to unzip and the flower explodes open zygomorphically.[6] The stigma and anthers often swing out towards the bird, enhancing the effectiveness of pollination.

The advantage of such as strategy from the plant's perspective is that it prevents nectar thieves and pollen robbers from entering the flowers, though some bees do seem to be able to open the buds.

Bird pollination also occurs in some Australian mistletoes in the genus *Amyema*. Large plants can produce enough nectar from their many flowers to make it profitable enough for larger honeyeaters to set up territories and aggressively drive off other birds. For a long time it was suggested that the leaves of these mistletoes mimic those of the host tree, as a strategy to reduce damage by browsing herbivores. The latest studies, however, conclude that this is not the case.

Relationships between mistletoes and birds such as flowerpeckers are some of the earliest bird–flower interactions described from Asia, as we'll see in Chapter 15. Indian naturalists have taken a special interest in this relationship, among them the remarkable ornithologist Sálim Moizuddin Abdul Ali (1896–1987), who documented his observations in the 1930s.[7] Fifty years later, Priya Davidar studied mistletoes in southern India that are

---

6 'Zygomorphic' means that it only has one plane of symmetry, like a human face. Zygomorphic flowers are found in plant families such as the mints and sages (Lamiaceae), the peas and beans (Fabaceae) and many others. Zygomorphy can allow for more precise placement of pollen on an animal, in contrast to radially symmetrical flowers like wild roses or buttercups.

7 Sometimes referred to as 'the Birdman of India', Salím Ali led a remarkable and long life and made huge contributions to Indian natural history. His Wikipedia entry is well worth reading.

both pollinated by flowerpeckers and have their seeds dispersed by the same birds (another double mutualism, to add to the one I mentioned in Chapter 3). She pointed out that, unlike the mistletoes pollinated by other groups of birds, the species that are reliant on flowerpeckers possess two unique features. The flowers are greenish and inconspicuous, rather than conspicuously red or pink, and they are produced as the fruits from the previous flowering season begin to ripen. Priya concluded that the flowers may, in effect, be mimicking the fruit, which are also greenish. The advantage for the plant, it seems, is that they can offer a much smaller nectar reward if the birds are fooled into opening buds that they mistake for fruit.

This is just one example of how plant behaviour can manipulate bird behaviour. As we'll see in the next chapter, there are others, but to appreciate them we need to think about how birds perceive the world around them.

CHAPTER 8

Senses and sensitivities

The preceding chapters, about pollination syndromes and the characteristics of flowers that exploit birds to move their pollen around, may have appeared to be mainly about plants. But in fact they were just as much an exploration of *animal* senses and physiologies. That's because we assume it's the preferences and requirements of the pollinators that mould how flowers look and what rewards they offer. To some extent this is true, though, as I argued in *Pollinators & Pollination*, flowers are not passive players in this relationship: they possess behaviours that can manipulate their floral guests in ways that are more advantageous for plant than animal. In this chapter I focus more explicitly on the avian side of the equation. I want to explore more closely what an understanding of bird senses can tell us about the natural selection that they impose on plants, altering how flowers look and smell over evolutionary time. There's also a further consideration of avian physiology and how that shapes the rewards that are on offer, convincing the birds to return to the same types of flowers time and again.

As intelligent, warm-blooded vertebrates, we can appreciate how birds use their sensory capabilities to survive and how this is linked to their physiologies. Mammals and birds are only distantly related on the tree of life – their last common ancestor roamed the Carboniferous landscape more than 300 million years ago. Nonetheless, birds possess the full range of highly attuned senses with which we (as mammals) are familiar: vision, hearing, taste, smell and touch. In fact, bird vision is arguably better than ours because many of them have four types of cone cells in their eyes – i.e. they are tetrachromatic – which means that they can also see ultraviolet light. Humans and some other primates, in contrast, are trichromatic, whilst most mammals are dichromatic.

In addition to their superior vision, birds have a complex and adaptable nervous system to take full advantage of the signals that are coming in from the world around them, much like mammals. There the similarities with us end, however. Birds typically live lives that are faster, more visually attuned to the ultraviolet parts of the spectrum, and they have at least one additional sense which we can't conceive: the ability to detect the Earth's magnetic field. These similarities and differences to our own senses not only shape the lives of the birds, they have also affected the evolution of at least some of the flowers on which they depend.

Seeing is perceiving

Sight is the most important sense which allows flower-visiting birds to locate and exploit nectar or other floral rewards, and indeed many strictly bird-pollinated flowers are brightly coloured. Red is particularly associated with bird pollination, but in fact even 'ornithophilous' flowers, as defined by the syndrome, come in an array of rainbow hues. These include the blues of some bromeliads, yellow in the Tree Tobacco we will encounter in Chapter 17, orange kniphofias, and even green, for example in a Caribbean member of the mallow family studied by Beverly Rathcke. Although you often read statements about the association between bird pollination and red flowers, birds do not have an innate preference for this colour. When Klaus Lunau and colleagues looked at this experimentally they found that

> Hummingbird-pollinated red flowers are on average less UV reflective, and white hummingbird-pollinated flowers are more UV reflective than the same coloured bee-pollinated ones. In preference tests with artificial flowers, neotropical orchid bees prefer red UV-reflecting artificial flowers and white UV-nonreflecting flowers over red and white

> flowers with the opposite UV properties. By contrast, hummingbirds showed no preference for any colour in the same tests.

Klaus recently pointed out to me that since the work of Austrian botanist Otto Porsch in the 1920s, if not earlier, it has been known that there is no single 'bird colour' for flowers. Despite the publication of papers such as 'Zukunftsaufgaben der Vogelblumenforschung auf Grund neuesten Tatbestandes' ('Future tasks for bird flower research based on the latest facts') and 'Grellrot als Vogelblumenfarbe' ('Bright red as a bird flower colour'), these invitations to see beyond red were ignored for many years.

Bright hues contrast strongly against the background vegetation and allow birds to locate the flowers from a distance. They may also reinforce the birds' spatial memory and allow them to make repeat visits to the same patches of flowers in a predictable sequence, a behaviour that is referred to as 'traplining' after the similar behaviour of hunters checking their animal traps. The concept was widely discussed for decades in the hummingbird literature, but it was not until a study led by Maria Cristina Tello-Ramos was published in 2015 that there was direct experimental evidence of the phenomenon. Traplining is a more efficient way of locating and exploiting predictable food sources, such as flowers on plants that do not move from one day to the next. It may seem surprising that birds with brains that are often described as 'no larger than a grain of rice' can remember the location of flowering patches, and the routes between them, within three-dimensionally complex landscapes. Hummingbirds, however, allocate a great deal of their physiology to their brains, which account for more than 4% of their body mass, the largest such proportion in the bird world. In relation to body size, hummingbird brains are two and a half times bigger, on average, than those of typical galliforms such as chickens, quails, pheasants and brushturkeys.

Size is not everything, however. The number and density of brain cells (neurones) that are devoted to specific tasks is also important. In hummingbirds, a region called the nucleus lentiformis mesencephali (helpfully abbreviated to LM by brain specialists) is 2–5 times larger, in relation to brain size, than in other birds. The LM processes fast-arriving visual information and is thought to be the key to hummers' abilities to hover, rapidly change direction, and interpret a world that is moving speedily relative to their bodies.

As Lars Chittka and Jeremy Niven emphasised in a review called 'Are bigger brains better?', the answer is 'no' – a lot depends on how the brain is organised, the size and connectivity of neurones, and so forth. Smaller brains can be more effective if they are appropriately constructed.

Lars is a bee expert, and he and I share a fondness for challenging received wisdom about pollinators and flowers, as demonstrated by the title of a paper that he wrote back in the mid-1990s with Nick Waser called 'Why red flowers are not invisible to bees'. In it, Lars and Nick took aim at a 'pervasive idea among pollination biologists … that bees cannot see red flowers'. As they went on to point out, this is a notion that 'has led many researchers to assume that red coloration is an adaptation by which flowers exclude bees as visitors'. This, in turn, has been interpreted as a reason why many hummingbird-pollinated flowers are red – that it reduces the ability of bees to find these flowers and steal the pollen, as much as it provides a strong signal to birds. Although it's an appealing idea, like most simple suggestions in ecology and evolutionary biology, it's wrong.

Or at least, it's not the whole story. Red flowers are less conspicuous to bees, but they certainly are not invisible to them, and these insects will find and keep visiting them if they offer appropriate rewards. Spanish scientists Miguel Rodríguez-Gironés and Luis Santamaría have suggested that because bees take longer to find red flowers, the optimal foraging theory that we encountered in the previous chapter should mean that they will focus their attentions more on flowers that reflect shorter wavelengths, towards the blue and ultraviolet end of the spectrum. Birds, with their tetrachromatic visual capacities, have no such problem, they are equally capable of detecting flowers of any colour so long as they stand out from the background vegetation, and so will search for the flowers that are most rewarding in a community. If the bees have paid less attention to the red flowers, this could result in strong natural selection for red bird-pollinated blooms.

A Chinese team led by Zhe Chen recently pointed out that these two hypotheses – 'bee avoidance' and 'bird attraction' – have not been properly tested over a large geographic scale. To remedy this, the researchers assessed the reflectance spectra for 130 plant species with red flowers from different continents. Looking at how flowers reflect light at different wavelengths using a piece of equipment called a reflectance spectrometer gives us a less subjective assessment of their colour than is possible from a verbal description based on our own vision, such as 'this flower is bright red'. The physicist-turned-pollination-ecologist Casper van der Kooi, amongst others, has emphasised just how complex flower colouration actually is. As well as the biochemistry of the pigmentation, there are other factors, such as the surface structure of the petals and internal reflectance within the tissues, that affect a range of flower characteristics, including colour contrast, hue, saturation, brightness, glossiness, fluorescence, polarisation and iridescence. Not all of these phenomena are measured by a reflectance spectrometer. But analysing the output in relation to colour vision models that have been developed for bees, birds and other pollinators allows us at least to get a sense of how they perceive the flowers, taking into account their own visual sensitivities.

Back to those red bird-pollinated flowers. Zhe Chen and colleagues found that, as expected, the colour vision models indicate that birds find red flowers very conspicuous, bees much less so. However, there were strong geographic differences between the flowers. Those from Africa and Asia were more conspicuous to bees than those from the Americas. They concluded that there is evidence to support both the 'bee avoidance' and 'bird attraction' hypotheses, but that the relative importance of these varies depending on the type of bird, the type of flower, and where in the world they are found. There is, however, another aspect to this that needs to be considered. These authors categorised flowers according to rigid boundaries, as 'hummingbird-pollinated', 'bee-pollinated', 'butterfly-pollinated' and so on. But of course, as we have seen, many flowers are pollinated by a mix of these animals, and that mix can vary over time and space. Ten per cent of the red flowers in their study were excluded from some analyses because they are pollinated by more than just birds, so an added complication to an already intricate story remains to be assessed.

Vision is not the only sense at play for avian flower visitors, however, and in some respects smell is a more puzzling aspect of bird pollination.

A nose for a flower and a taste for insects

One of the oddities of many (though undoubtedly not all) bird-pollinated flowers is that they lack an odour that is perceptible to humans. This is often seized upon, especially in textbooks, as being an evolutionary response to birds having a poor sense of smell. If the birds are not using scent to locate flowers, goes the argument, it makes sense for the plant to not synthesise the complex odour bouquets of the kinds that attract other pollinators like bees and flies. The problem is that this is false reasoning. In fact many birds have well-developed olfactory systems – think of vultures, which find cadavers even in dense undergrowth. This has not always been appreciated; as a 2015 study by Jeremy Corfield and colleagues put it, 'early ornithologists debated if birds had a sense of smell at all'.

These early ornithologists included John James Audubon, whose 1826 'Account of the habits of the turkey buzzard … with the view of exploding the opinion generally entertained of its extraordinary power of smelling' influenced a certain Charles Robert Darwin to try his own experiments with Andean condors. Aside from the early experimental evidence, which as Darwin noted was mixed,[1] there was also the matter of anatomy and the fact that the olfactory bulb (OB) in the brains of birds is often relatively small. The OB, of course, is fundamental to the perception of different smells. However, the ratio of OB size to that of the rest

1 The keen-eyed reader will notice that throughout the book I've tended to refer to scientists by their forenames, because first and foremost they are people, with names and lives of their own. When it comes to historical figures such as Charles Darwin, however, I have made an exception: calling him 'Charles' feels simply wrong. See also Gilbert White in Chapter 13, amongst others.

of the brain varies enormously between different groups of birds and seems to have evolved to reflect different ecologies such as food type, migration, finding a mate and so forth. Hummingbirds and parrots certainly have relatively small OBs, but other flower-feeding birds have hardly been studied.

As I noted in the discussion of bird vision, size is not everything, but the textbook fallacy that birds do not use scent to find flowers has rarely been tested. In fact, I am aware of just three studies that experimentally asked this question, and all used hummingbird visits to artificial sugar-water feeders as their setup, and tested whether the birds could discriminate between rewarding feeders that were scented and scentless unrewarding feeders. In two of the three studies (involving four out of five species) the birds were able to discriminate between the feeders that were scented and those that were not. This is a small sample, and more work needs to be done on a diversity of species, but it at least shows that some hummingbirds can find flowers using their sense of smell.

But if flower-visiting birds can smell the targets of their foraging, why are so many bird-pollinated flowers lacking in odour? There are several possibilities, which are not mutually exclusive. Producing the cocktail of chemicals that give a flower its distinctive scent, which can number in the tens or hundreds of different types of molecules, is expensive in terms of energy and resources that could otherwise be used for growing, making more flowers or provisioning seeds. Any plant that can forgo the expense and attract its pollinators solely by visual means could be at an evolutionary advantage. However, flowers may have evolved this strategy of olfactory inconspicuousness to make them less obvious to insects that would otherwise raid the flowers for pollen and nectar. Cornell University biologist Rob Raguso has done more analysis of flower scents than almost anyone else I know, and in a recent email exchange he pointed out that it's certainly not true that all flowers that are pollinated by hummingbirds are scentless. Together with Danish researcher Jette Knudsen, Rob collected samples from 17 species of hummingbird-pollinated flowers in Ecuador. In eight of them they found traces of volatiles that they speculate could 'attract alternative pollinators, repel enemies or represent vestiges of a scented ancestry'. Of course they might also be involved in attracting the birds. Such scents may also be more widespread amongst bird-pollinated flowers, because, as Rob noted,

> some of the most prevalent compounds, such as sesquiterpenes, have very high thresholds of perception for most human noses … [and] … the weak scent in many bird-pollinated flowers may indicate that the nectar itself is the source of the scent.

In other words, the flowers are scented, it's just that we humans are barely able to detect the odour. Just as an extra set of visual cones means that birds see the world in a rather different way to us, it's entirely possible that they smell it differently too.

What about taste? Where does this fit into flower feeding by birds? In some ways this is even more difficult to study, because there is no obvious structure in the brain that deals with taste, no gustatory equivalent to the OB. Instead, to understand how birds perceive taste requires a study of taste receptors in the nervous system. In 2021 an international team discovered that an evolutionary switch in the taste receptors which detect savoury umami flavours to sweet, sugary flavours seems to have occurred early on within the evolution of the songbirds. As the authors phrased it:

> This ancient change facilitated sugar detection not just in nectar feeding birds, but also across the songbird group, and in a way that was different from, though convergent with, that in hummingbirds.

The team concluded that this evolution occurred about 30 million years ago, before the first ancestors of songbirds moved out of Australia, where the group originated. This fits quite neatly with some of the ideas that Tim Low puts forward in his book *Where Song Began: Australia's Birds and How They Changed the World*, and particularly the importance of nectar for these early songbirds.

In his words, it was a 'food worth defending' in the harsh and unproductive landscape of that ancient continent.

As well as being able to taste the sugars in nectar, flower-visiting birds also have a taste for other types of food – even hummingbirds, arguably the most specialised of the avian nectar feeders. If we consider the question 'what do hummingbirds eat?', the obvious answer is 'nectar', of course. But nectar does not make for an especially nutritious diet. As I described in Chapter 6, it's mainly water plus sugars (in varying concentrations) plus some amino acids and a few secondary compounds such as alkaloids, and often yeasts and other microbes. No animal could subsist for its entire life on just nectar, not least because it's impossible to build bones and muscles from such a limited diet. Even insects like some butterflies, which appear to survive only on nectar, gain most of their body-forming nutrition from the plant materials that they consume as caterpillars, and as adults will often seek out inorganic nutrients from dung, urine and mud. Yet the myth persists that hummingbirds have a diet that comprises nectar almost exclusively. I have seen this repeated in numerous books on these birds, some quite recently. But it's wrong.

In fact we have known for a long time that hummingbirds feed on arthropods such as insects and spiders, and that these comprise a significant proportion of the birds' diets, especially in the Neotropical dry season when few flowers are available. For example, Brigitte Poulin and colleagues, working in dry, seasonal vegetation of north-eastern Venezuela in the 1980s concluded that the hummingbirds which they studied 'had low nectar intakes and … [fed] … extensively on small soft-bodied arthropods'. Similarly, in 1971, Allen Young published a study called 'Foraging for insects by a tropical hummingbird' in which he showed that in Costa Rica, Long-tailed Hermits spent significant amounts of time gleaning insects from leaves and even from the webs of orb-spinning spiders, concluding that during some parts of the year this species of hummingbird

> may devote the bulk of its foraging effort to
> searching for insects, while at other times, the
> habit may be primarily nectarivorous.

Since then, numerous other studies have shown the same thing: hummingbirds eat insects, often frequently and in large amounts. These studies were not the first to show this, and in fact Allen Young cited Helmuth Wagner, whose own studies, published in 1946, showed that

> The food of hummingbirds is determined primarily by habitat and season. A given species may feed mainly on nectar or mainly on insects, depending on the time of year. The majority of the hummingbirds found in Mexico are not dependent on flowers, their migrations being determined by food supply in general rather than by the supply of a particular kind of food. When flowers are lacking, or when the food they provide is inadequate, hummingbirds live on insects and other small animal life.

That wasn't even the start of our understanding of the complexity of hummingbird feeding habits – we can go back even further in time, because Wagner cites work from as early as 1916 and emphasises the distinction between conclusions drawn from 'merely casual observations … [and] … very careful investigations'.

Until recently, understanding the arthropod component of hummingbird diets was time-consuming, complex and potentially damaging to the birds. It required that researchers spend days making observations in the field, examining regurgitated pellets, or taking a close look at the gut contents of birds, either deceased or living. Checking what is in the gut of a live hummingbird used to rely on the invasive technique of feeding the bird an emetic substance and then collecting the vomit. Nice. More recently a non-invasive technique using next-generation DNA sequencing and metabarcoding has been developed by researchers Alison Moran, Sean Prosser and Jonathan Moran. This means that arthropods can be identified from the faecal pellets of nestlings, collected from nests after they have fledged. When they applied this approach to Rufous Hummingbirds these researchers found DNA belonging to three different classes, eight orders, 48 families and 87 genera of arthropods. One pellet alone contained the molecular evidence of 15 different families.

Not only does a study of the detailed ecology of hummingbirds demonstrate that the 'nectar only' idea is a myth, but anatomical investigations have shown that several hummingbird species (and by implication, perhaps all of them) have evolved a beak with unique adaptations to catching insects on the wing. Again, this should not surprise us as, depending on the species, the time of year, and whether the birds are breeding, hummingbirds can spend up to 70% of their food-seeking time foraging for insects.

There are also numerous reports of hummingbirds feeding on other nutritious fluids, including plant sap and the honeydew produced by some insects, as well as the juice (and possibly pulp) from fruits. In the latter case it seems unlikely that they are contributing to seed dispersal, but I wouldn't rule it out – there are always surprises awaiting us just around the corner.

If I'm labouring the point about what hummingbirds eat, it's to emphasise that not only are the myths about hummingbird feeding incorrect, but that repeating them up until the present day is due to either repetition of what others have stated, or because when birders look at hummingbird behaviour and photograph their activities, it's often when they are feeding on flowers or artificial sugar-water feeders. But there's so much more to these wonderful birds than this, and no birds subsist solely on nectar, no matter how specialised they may be at exploiting flowers. Even that most specialised of nectar feeders, the Sword-billed Hummingbird that we'll hear more about in Chapter 11, takes insects from flowers to supplement its diet.

What about the other major groups of nectarivorous birds, such as sunbirds, honeyeaters and white-eyes? How varied are their diets? The answer is that they are even more diverse than the hummingbirds. In a paper published in 1980, Graham Pyke reviewed what was then known about the diet of honeyeaters. He showed that many honeyeaters (especially the bigger species) frequently include a large proportion of fruit in their diets, and that some such as the Painted Honeyeater are, on balance, more frugivorous than nectarivorous. Arthropods are often consumed and, again, for species such as the Green-backed Honeyeater, these make up the majority of their diet. Intriguingly, the latter species is dully coloured brown and yellow, and except for a slightly longer beak, looks more like a species of leaf warbler than a vividly hued honeyeater.

In Chapter 6 I noted that pollen can be tricky to digest for some avian groups, especially for hummingbirds. However, it features in the diet of some honeyeaters, parrots and Galápagos finches, all true omnivores for which nectar is only part of a balanced diet. What flower-visiting birds eat, however, is much influenced by the time of year that they are actively foraging. Birds typically follow the cycles of food availability, which change seasonally, even in the tropics where, rather than having four more-or-less distinct seasons, as occurs in the temperate zones, there may only be two, a wet and dry, or a wet and a very wet!

Timing is everything

A few years ago, in an antique shop in Northampton, I came across an oil painting by the South African artist Henry Bredenkamp. I saw it from a distance, and what initially caught my eye were not the ochres, yellows and browns of what was clearly a semi-arid landscape, but the scattered patches of red that were the impressions of tree aloe flowers. In the dry winter of western South Africa, these plants are amongst the few to be flowering, providing nectar for birds (and sometimes insects) that depend on this regular annual bounty. I've seen similar patterns in other parts of the world, such as the so-called 'Indian paintbrushes' (*Castilleja* species) that light up the dry hills of the southwestern USA (Plate 11) and attract migratory hummingbirds.[2] Flowering at a time when other floral

---

2 Interestingly, in a recent study Evan Hilpman and Jeremiah Busch at Washington State University found little relationship between flower traits and the main flower visitors, and concluded that '*a priori* notions of pollination syndromes in this system are overly simplistic and fail to predict which animals most frequently visit *Castilleja* in natural populations'.

resources are scarce is thus one strategy that bird-pollinated flowers can adopt to ensure frequent avian visits.

Across the globe, many flower-visiting birds are migratory. We see this in northern Europe where, each spring, some warblers arrive with 'pollen horns' on the base of their bills, symbolic of the flowers that have recharged their energy-consuming flights from Africa. Mirroring this, New World warblers such as the Tennessee Warbler travel long distances to winter in the warmth of Central America. Arthur Cleveland Bent's 1953 *Life Histories of North American Wood Warblers* quotes an account from ornithologist Alexander Skutch in Costa Rica, that this species is 'fond of flowers, especially the clustered heads of small florets of the Compositae and Mimosaceae, and of the introduced Grevillea that sometimes shades the coffee plantations'.

In Asia, some of the northern populations of Swinhoe's White-eye are likewise migratory, in the winter moving from eastern China down to Myanmar, Thailand and other warmer parts of the region. The *Birds of the World* account of this species describes 'large flocks found during last days of Oct[ober] in lower Yangtze basin, apparently preparing to move [south]'.

African sunbirds are not currently thought to be such long-distance migrants, though outside of the breeding season they will certainly move shorter distances in order to find nectar. In Australia, some honeyeaters make nomadic journeys of hundreds of kilometres in search of seasonal nectar sources, and there are online accounts of tens of thousands of birds moving between Queensland and New South Wales.

Even in the tropical rainforests of Amazonia, where temperature and daylength hardly vary, large changes in rainfall between seasons result in local fluctuations in flower and insect abundance. Ecologist Peter Cotton commented that 'these seasonal fluctuations in resource abundance correlated very closely with variation in hummingbird species richness, hummingbird abundance and the total biomass of hummingbirds in the assemblage'.

Back in North America, the best-known bird pollinator migrations are, of course, undertaken by species such as the Rufous Hummingbird, which each year travels from central Mexico as far north as southern Alaska, and back. These movements are tracked by citizen science projects such as Journey North (journeynorth.org), amassing huge amounts of data. It is on this sort of epic aerial expedition that migratory birds bring into play their abilities to sense the Earth's magnetic field.

Perhaps less well appreciated is the fact that some species of hummingbird-pollinated plants time their flowering to coincide with the arrival of the birds. This was elegantly demonstrated by Nick Waser's study of the consistent flowering of Ocotillo, a semi-succulent desert shrub. At his sites near Tucson, Arizona, Nick found that the flowering was timed to the arrival of hummingbirds between the end of March and mid-May. Ocotillo is a major nectar source for species such as Black-chinned, Broad-tailed, Costa's, and Broad-billed Hummingbirds in southwestern North America, fuelling their local breeding or onward journeys, and in return (as Nick also showed) receiving pollination services.

Ocotillo is not the only source of nectar for these birds, of course, and the flowers of the plant are exploited by other animals, including some bees. Together these interactions form a local web of relationships that extend outwards and link birds and their flowers across time and space. That's where our story next takes us, into the knotty world of ecological networks.

CHAPTER 9

Codependent connections

Ecologists like myself who are interested in describing and understanding the interactions between flowers and pollinators within ecosystems have traditionally used a descriptive, cataloguing approach. Patient observations of the visitors to flowers yielded lists of the likely pollinators, and these data could be tabulated or graphed to show the patterns. Typically, this took the form of '*n* species are bee-pollinated', '*n* species are bird-pollinated', and so forth, often with a comparison of the different flower sizes, shapes and colours. Such an approach might appear rather old-fashioned, but it provides us with important, basic information about the ecological connections within a community of plants and animals and is not to be dismissed lightly. Indeed, observations collected in this way by early-twentieth-century naturalists still find their way into modern analyses. For example, the Interaction Web Database (www.ecologia.ib.usp.br/iwdb) includes a set of observations from Pikes Peak in Colorado collected by Frederic Edward Clements (1874–1945) and Frances Louise Long (1885–1946), and originally published in their 1923 book *Experimental Pollination: an Outline of the Ecology of Flowers and Insects*.[1]

Over the past 20 years or so, ecologists have increasingly turned to more sophisticated ways of analysing their own and these inherited data, using network approaches that were originally developed for other disciplines such as physics and computer sciences. These take a range of forms, including 'bipartite' graphs that visually show the connections between two parties, in this case plants

1 Despite the title, at least one avian flower visitor features in the data, the Broad-tailed Hummingbird.

and pollinators, and various metrics that describe the structure of the interaction data. Such powerful statistical approaches provide important insights into what has been described by Spanish ecologists Jordi Bascompte and Pedro Jordano as the 'architecture' of species interactions.

It is possible to look at all of the pollinating relationships within a plant community, including both insects and birds, and I've worked with some very talented ecologists on analysing such networks. Among them are Ruben Alarcón and Stella Watts, who both worked on these networks as part of their PhD studies. A paper Ruben wrote in 2008 with Nick Waser and myself, entitled 'Year-to-year variation in the topology of a plant–pollinator interaction network', was one of the first to document how the links between pollinators, including insects and hummingbirds, and the flowers that they visit, change between years even in the same community with largely the same species present. Stella's work in Peru, some of which I described in Chapter 7, included an assessment of how such networks varied across a sequence of parallel Andean valleys.

Complete network studies like those undertaken by Ruben and Stella are vital, but they can be overwhelming in their complexity and logistical requirements. For that reason, many researchers focus on just a part of the network, constrained by taxonomy, for example a particular plant family or group of pollinators. One of my earliest papers to adopt this approach, at least in part, looked at the pollinators of a small group of co-occurring asclepiads in the grasslands of KwaZulu-Natal, South Africa – but none of them, unfortunately, was bird-pollinated.

Reversing this approach, I've also been part of work that focused just on the birds and their interactions with flowers within a community, collaborating with Danish and Brazilian ecologists such as Bo Dalsgaard, André Rodrigo Rech and others. Much of this has used hummingbirds as a model system for understanding the structural complexity of plant–pollinator networks, and how they vary in space and time, and according to factors such as climate. Less bird–flower network research has been done with groups such as sunbirds and white-eyes, though this is changing thanks to the efforts of African and Asian ecologists. Another important gap in our understanding is that few of these studies include birds from other families, even though, as we've seen in the earlier chapters, these are important flower visitors.

How plant–pollinator networks are structured

The relationships between flowers and their visitors in networks are never random; they always have their own structure, an internal logic that is particular to the time and place where they were studied. Not only that, but there are consistent aspects of that structure which transcend time and space, whether one is studying honeyeaters in Australia, hummingbirds in South America, or sunbirds in Africa and Asia.

One way to visualise these networks, and their consistencies, is to think about a child's set of wooden blocks and the box in which they are housed, just large enough to contain those blocks in a single layer, no more and no less. The blocks do not represent species, but the potential or realised interactions between species. The species themselves are found along the edges of the box: the pollinators down the left side, the plants along the top. One block is one interaction between a plant and a pollinator, so if our box contains 50 blocks, we could represent a network of, say, 10 pollinators and 5 plants. That would be a very small network: a box containing as many as 1,000 blocks (say 50 pollinators and 20 plants) is quite routine. The thing is, not all of those blocks represent actual interactions: as I said above, the interactions may be potential or realised. If we painted the blocks according to whether or not a given pollinator interacts with a given plant (say, blue for the interactions which do occur and white for those that don't), then something interesting would emerge: most of the blocks would be white, with only a minority blue. That's because, in a real ecological community, most of the pollinators and flowers do not interact. This might be due to a mismatch in the morphologies of the animals and plants, or because the animals make active choices about which flowers are most rewarding to them, or because the plants bloom at a time when some of the potential pollinators are not active. In a network of 1,000 possible interactions, it's not unusual for 70% or 80% of those interactions to not occur.

If our plants and pollinators were both arranged alphabetically, then we'd be presented with a random scatter of occasional blue blocks in a sea of white. But nature doesn't understand our alphabet, and there are much more appropriate ways to order those species in a way that tells us something about their ecology. One obvious way is to rank them according to how many partners they have, from the most generalist to the most specialist. If we do that for both plants and pollinators, structure begins to appear (Plate 13a). The first row and the first column may be full of blue blocks, showing that the most generalist plant is exploiting all of the available pollinators, and the most generalist pollinator is visiting the flowers of all of the available plants. The next columns and rows in sequence will each have one or two fewer blue blocks, until we get to the final few columns and rows that have just one blue block in them: pollinators that specialise on one plant, which is often the most generalised species, and plants that possess just one pollinator, again usually one of the generalists.

Step back and look at the pattern that you have created and you'll see that a more-or-less triangular shape has emerged in which generalists interact both with other generalists and with specialists, whilst specialists largely interact just with the generalists. In reality it's never a perfect triangular pattern, and there are often white and blue blocks in places we might not expect them. But if we test the pattern statistically we usually find that it's significantly different from a random set of interactions. We term this pattern 'nestedness', or speak of a nested pattern of interactions, and it turns up with amazing regularity in plant–pollinator networks.

As I was working on this chapter and thinking about a suitable title, I discussed the idea of calling it 'Codependent connections' with my wife, Karin Blak, who trained as a therapist and wrote *The Essential Companion to Talking Therapy*. Karin's specialism is relationships, though she no longer practises and now works as an author; she's my go-to person when it comes to reflecting on my writing, and we often discuss the parallels between human and ecological relationships. Karin pointed out that 'codependent' did not necessarily mean that each partner was equally dependent upon the other (as I had naively assumed). Often there's a degree of asymmetry in codependent human interactions, and that's precisely what nestedness is demonstrating. A plant often relies on the services of a bird that, in turn, visits and pollinates a wide range of other flowers. Conversely, a pollinator may be dependent on a plant that recruits a wide range of other pollen vectors. This is especially true of specialist bees that take their pollen only from one genus or family of plants, if there's only a single representative of that family within a community. Birds are less specialised in this sense, but it may be that an avian flower visitor relies on the nectar from just one plant in a locality, particularly at a time of the year when few others are in bloom.

There are other patterns within our box of blocks, however, and to see them we need to repaint the blue blocks with a wider palette of colours. Let's paint the bee–flower interactions yellow, the bird–flower interactions green, the fly–flower interactions orange, and so forth. Let's then arrange the newly coloured blocks in a different way, ranking the species not from most generalist to most specialist, as we did before, but arranging them so that the birds are grouped in consecutive rows and all of the bird-pollinated flowers are grouped in adjacent columns. We'll do the same for the other types of pollinator and their respective flowers, and we will see another pattern emerging. Along the axis of blocks going from top left to bottom right we see clusters of blocks of the same colour (Plate 13b). We refer to this as a modular structure, and again it's usually not perfect, in that some interactions will be out of place, but clear enough to be different from a random pattern.

Things can get even more complicated when we take out individual modules and consider just the interactions between that set of pollinators and the plants they service. Very often we see a nested structure within the module. That's why looking at a restricted set of pollinators, such as just the birds, is an appropriate approach to managing the complexity of whole communities: interactions at this smaller scale reflect what's happening at the larger scale.

These kinds of considerations about plant–pollinator networks can seem rather abstract and divorced from the natural history of what is happening in real communities of organisms. So let me give you some examples of real networks.

Real birds and real flowers in networks

Since the early 2000s, Bo Dalsgaard has worked in the Caribbean, trying to understand why the networks of flowers and hummingbirds are structured in the way that they are. Amongst the variables he's considered are the kinds of hummingbirds that are found on particular islands, the size and type of island, the elevation, and current and past climates. It's painstaking work involving many hundreds of hours in the field, watching patches of plants and recording the visits of birds to flowers. Although camera traps are being used with increasing frequency to record the sometimes very infrequent visits to flowers, there's a resource limit to their use: how much money do I have and how many cameras can I afford to deploy? In a study on which I collaborated with Bo, published in 2008, we used a network approach to investigate whether hummingbirds visit an exclusive subset of flowers that provide resources which are unobtainable to insects. We found that this was true for only seven of the 36 plants in the network. Not only that, but the smallest bird in our study, the Antillean Crested Hummingbird, which typically weighs around 3 grams, behaved more like the insects in its choice of flowers. Larger bird species, such as the Rufous-breasted Hermit and the Purple-throated Carib (which weigh up to 8 and 10 grams respectively) were more 'functionally specialised' in the sense that they visited flowers which insects avoided. There might be several reasons for this, including the possibility that the larger birds aggressively exclude the smaller birds from their preferred flowers. However it's also likely that small body size allows some birds to gain enough energy from flowers that produce relatively little nectar, flowers that originally evolved to reward just insects.

More recently, on the other side of the Atlantic, a joint Czech–Cameroonian team led by Štěpán Janeček from Charles University has been studying sunbird–plant networks on Mount Cameroon, an active volcano and the highest peak in western and central Africa. The researchers used both automated video recording and direct observation to document bird–flower interactions in four different forest types that dominate at different elevations of Mount Cameroon, and which vary in the plants that flower in the dry and wet seasons. They found that whilst the networks as a whole were highly generalised, with sunbirds and flowers frequently sharing partners, this varied seasonally and with elevation. In lower-elevation forests, and in the wet season, higher rates of specialisation were observed, while in the dry season, when fewer other plants were in flower, trees were rather generalist and enjoyed visits by multiple species of sunbirds.

Sufficient networks of this kind have now been documented that we're at the point where we can start to make comparisons across biogeographic regions, to assess the similarities and differences between networks composed of different groups of birds and plants. In 2017 Brazilian ecologist Thais Zanata combed the published literature and sought out unpublished data, bringing together 79 hummingbird, nine sunbird and 33 honeyeater networks in order to test the hypothesis that hummingbird–flower networks were more specialised than those of sunbirds, and in turn that the honeyeaters were the most generalist flower visitors. This turned out to be partly true. The hummingbirds and their flowers were more specialised than the honeyeaters. The sunbirds, however, were statistically no different to the other groups in their level of specialisation. This might be due largely to the low sample size of sunbird networks, all of which were from Africa. For such a diverse and spectacular group of birds, we know surprisingly little about their ecologies. In the future I would also like to see comparisons made with the other types of flower-visiting birds, because we know even less about how these fit into the bird–flower networks.

Until recently such network studies simply documented the relationships in terms of which bird visits which flower, not the outcome of the interactions in terms of pollen removal from anthers and deposition on stigmas. As you might imagine, this is difficult to do at a community scale, but ecologists are starting to take up the challenge. On the Mambilla Plateau in southeast Nigeria, for example, Charles Nsor, William Godsoe and Hazel Chapman have studied the amount of pollen being transported by sunbirds in an Afromontane forest habitat. As with other sunbird studies, they found significant levels of generalisation in these birds, compared with hummingbird networks. One of the consequences of generalist flowers being visited by birds that visit multiple other types of flowers is that the wrong pollen can end up on a stigma. We're just beginning to understand the consequences of such pollen–stigma mismatches in these bird–flower networks.

Plant ecology is sometimes messy

Plant species usually don't grow alone, but typically live in diverse communities with other species. As we saw above, it is not unusual for some of these plants to share pollinators. If they do this, and they flower at the same time, the receptive female surface of the flower – the stigma – can receive the pollen from different types of plants as well from individuals of their own species. We term this 'heterospecific' pollen deposition, and there has been much discussion amongst ecologists about what the consequences of this might be for plant reproduction. One consequence, of course, is that any pollen arriving on the wrong stigma is wasted as far as fatherhood is concerned unless, under rare conditions, interspecific hybrids result. The second consequence, for the flower receiving the wrong pollen, is that it may clog the stigmas and prevent the correct – 'conspecific' – pollen from gaining a foothold. The potential for this to decrease seed production has been considered theoretically, but relatively little has yet been published on this phenomenon and its impact on reproduction, particularly in highly diverse tropical communities across different seasons. It was a pleasure, therefore, to work with my Brazilian colleagues on a study led by Sabrina Aparecida Lopes, in which we looked at heterospecific pollen deposition (HPD) in a Campo Rupestre ('rocky grasslands') community in the Serra do Espinhaço, comprising 31 hummingbird-pollinated plant species and nine attendant hummingbirds.

One of the novel things that the study addressed was the question of how changes in the community of co-flowering plants over the course of the season affected patterns of HPD. By sampling the stigmas of the plants at different time periods and also looking at fruit and seed set, Sabrina's team established that there's significant seasonality in HPD: those plants that flower in the dry season had a much greater diversity of pollen on their stigmas, and a greater proportion of stigmas with heterospecific pollen, in comparison to plants that flowered in the wet season. Perhaps not surprisingly, the stigmas of those plants that were more ecologically generalised, and visited by a greater diversity of hummingbirds, encountered more HPD.

What is much more surprising, however, and contradicts the prevailing view of HPD, is that the relationship between HPD and seed set was *positive*: flowers which received more of the *wrong* pollen tended to have *greater* reproductive success in terms of seed set. We're not for one moment thinking that this is a matter of cause and effect, but rather that there are benefits to sharing pollinators with other flowers, and being more generalised, even if a side effect is that there is more HPD. By coincidence, the very month that we published that work a study in a high-Andean plant community in Argentina by another Sabrina (Gavini) and colleagues demonstrated exactly the same thing. Their research considered all pollinators, not just birds, but it does show that our results were not an isolated case. It also adds another dimension to the importance of conserving plant diversity in natural communities.

The study of bird–flower networks can be used to address fundamental questions about the ecology and evolution of these interactions. The networks also have a more applied purpose, however, when it comes to understanding how they can remain stable and resilient in the face of human disturbance to habitats, climate change, and other assaults on the natural world. The Swiss team of François Duchenne, Rafael Wüest and Catherine Graham, for instance, recently showed that the way in which hummingbird–flower relationships are organised seasonally affects their stability. This implies that the loss of certain species, for example those that flower or pollinate at particular times of the year, may profoundly affect the local community. In order to test such ideas, other hummingbird–flower researchers have started to take an experimental approach to understanding networks. Kara Leimberger, Adam Hadley and Matthew Betts removed all of the heliconia flowers within forest fragments at Las Cruces Biological Station in Costa Rica in order to simulate the local extinction of an important nectar resource. They expected that the birds would to some extent 'rewire' their interactions and forage more often on alternative nectar sources. That did not happen, which I find interesting as it suggests that hummingbirds are less flexible in their diets under some circumstances. In the future I hope that we'll see more of these sorts of experiments, perhaps using state-of-the-art molecular biology techniques such as swabbing flowers for the DNA left behind by bird visitors, as recently pioneered by Knud Jønsson from the University of Copenhagen.

Of course, pollinating birds and insects share their plants with a host of other organisms that occur in these communities, including herbivores and parasites, amongst others, all of which are part of a wider network of ecological actors. Some of these additional species also make a living on and within flowers, and occasionally birds, with at times alarming consequences for both pollinator and pollination. These are the critters that we'll consider in the next chapter.

CHAPTER 10

Hitchhikers, drunks and killers

Arriving at our field site in Guyana, we walked the last couple of kilometres from the end of the dirt road to the riverside where our hosts had suggested that we camp. The transition from savanna to rainforest was a gradual one, and as we trekked through the ecotone, trees and shrubs adapted to the open, drier grasslands were replaced by species which needed more moisture and shade.[1] I was momentarily distracted as a hummingbird flashed past, and I made the mistake of trying to follow the bird with my eyes while still walking. Suddenly I felt a powerful hand on my chest and I was forced to stop. Peter, our guide, was pointing at the trail in front. It took me a moment to realise what he had seen – across our path lay a large snake. 'Bushmaster,' said Peter, calmly. I was not quite as calm. Just two more steps and I would have had a close encounter with one of South America's most dangerous serpents. Picking up a twig, Peter threw it gently at the snake, which struck before the twig even reached the ground, then slowly moved off into the undergrowth. Without his keen eyes our field work in Guyana might have come to an abrupt conclusion, and after that morning my flower and bird observations were made with one eye forever on the route in front of me.

This incident was a sobering reminder that the networks of birds and flowers that I discussed in the previous chapter do not sit

---

1 An ecotone is a transitional area from one type of habitat to another. It's quite often richer in species than either of the adjacent areas, because species from both habitats can find food, nesting sites or other resources there.

in isolation within an ecosystem. A host of other organisms share their space and sometimes interact with both the avian and botanical partners, occasionally in subtle and extraordinary ways. These species range from the microscopic (yeast, bacteria), through the mesoscopic (tiny mites that can just about be seen with the unaided eye), up to the macroscopic herbivores and predators that feed on flowers, seeds and birds. The ecological nature of these relationships can be classified as predatory, parasitic, commensal or mutualistic, depending on the outcome for the different partners: does it have a positive (+), negative (–) or neutral (0) effect on their lives?

Ecologists use this +, –, 0 notation as a shorthand for these outcomes. For example, most of the relationships between birds and flowers that we are dealing with in this book have a positive outcome for both species involved: the plant gets pollinated, and the bird receives a food reward. We define this as a ++ ('plus-plus') relationship and describe it as mutualistic. Parasitic and predatory interactions are considered +– ('plus-minus') because one species benefits at the expense of the other. Commensal relationships (+0 or 'plus-zero'), in which one party benefits and the other is neither helped nor harmed, are common in nature. They include interactions such as mosses and lichens living on the trunk of a tree, birds nesting in those trees, or indeed the wide variety of invertebrates that live harmlessly with birds in their nests. In fact commensalism is so ubiquitous that it's often considered less frequently than the other, high-profile interactions, especially those that demonstrate 'nature, red in tooth and claw'.[2]

When it comes to birds and flowers, the range of other species involved in their interactions is astonishingly diverse. As the title

2 The famous phrase is from the 1850 poem 'In Memoriam A.H.H.' by Alfred, Lord Tennyson. The section of the poem that includes this phrase refers a few lines later to dinosaurs (before that word was coined) as 'Dragons of the prime, / That tare each other in their slime'. Although this was written almost a decade before Wallace and Darwin presented their views on the evolution of life on Earth, later commentators often link 'red in tooth and claw' with the evolutionary struggle for existence, 'survival of the fittest', and it is even used as an excuse for the worst excesses of human beastliness. Not all of nature is like this; many relationships between species are less red and involve neither teeth nor claws.

of this chapter hints, the outcomes that these interlopers bring to bird–flower relationships are likewise varied.

Fermentation in a flower

Australian ecologist Michael Whitehead has an interesting and unusual hobby: he carefully cultivates the yeasts he finds in flowers and uses them to brew beer. Depending on the type of yeast he finds, his beers taste floral or repugnant – it's a little hit and miss, as Michael has documented on his blog.

Although we associate the word 'yeast' with various species of *Saccharomyces* that are used in baking and brewing, in fact it does not refer to a single species or even group of species. Yeast is the collective term for any fungus that is largely single-celled during its life cycle. At least 1,500 species of fungi with a yeast morphology have been described, and many, many more await discovery. A lot of yeasts are associated with plants and are particularly abundant within flowers, where they feed upon, and often ferment, the sugars within the nectar. This can have consequences for floral relationships with pollinators. One is that the amount of sugar available to the pollinators will be lessened if the yeasts have metabolised it as they grow and reproduce. Related to this, the chemical by-products of yeast activity can accumulate in the nectar, changing its odour, taste and nutritional value. These by-products include ethanol and other alcohols produced by fermentation of the nectar sugars, and any pollinators drinking such a brew can experience the typical effects of drunkenness. This has best been documented in bees, but is also known to occur in hummingbirds and is one reason why people are encouraged to regularly clean their sugar-water feeders. Drunk pollinators sound amusing, but as several videos on YouTube document, for hummingbirds it results in an inability to fly properly, sometimes leading to collisions with windows and cars. When it happens in the wild, of course, it can then affect the ability of birds to position themselves in ways that allow them to effectively access the nectar, which is usually also the best position for picking up and depositing pollen.

One obvious question is: how do the yeasts get into the flowers in the first place? It appears that there are two routes. The first is that airborne yeast spores, which are common in the atmosphere, fall passively into the open blooms. This is likely to happen infrequently: a joint Mexican–Spanish survey led by Carlos Herrera found that, depending on the region, between less than a third and just over a half of the species had yeasts in their flowers. If passive colonisation was the main route by which yeasts found their way into flowers, I think we'd expect these proportions to be higher. The second route is that the yeasts are transferred on the bodies of flower visitors, especially the mouthparts which encounter that all-important nectar. This is likely to be a much more important mechanism of colonisation, which is supported by studies of flowers within a community that show a close agreement between the types of pollinators and the types of yeasts. The bird-pollinated flowers of Tenerife that I will describe in Chapter 12, for example, contain a community of yeasts that are distinct from the yeasts found in bumblebee-pollinated flowers, which in turn differ from the yeasts in flowers visited by the non-social 'solitary' bees. Nectar sugar differences may account for some of this variation, but the conclusion of the study by Moritz Mittelbach and others was that it was mainly due to inoculation effects: birds inoculate bird flowers, bumblebees inoculate bumblebee flowers, and so on. Further evidence that it's the pollinators that largely determine which flowers contain particular yeasts comes from work on a hummingbird-pollinated monkeyflower in California. In this species, the spatial distribution of a single type of yeast is correlated with the foraging patterns of the birds, likely a consequence of traplining or territoriality. When the researchers sampled the birds' bills and tongues, by getting them to drink from feeders containing sterile sugar water, they discovered that they were carrying viable yeast cells.

It's not just yeasts that can colonise flowers, of course. Bacteria make their way into these floral environments, probably also via birds, but their role in floral ecology is less well understood. However, it is known that bacteria differ from yeasts in their effects on nectar in some fundamental ways. This was demonstrated by a 2018 study of a hummingbird-pollinated shrub in which biologists

Rachel Vannette and Tadashi Fukami inoculated flowers with either a yeast or a bacterium. Rachel and Tadashi discovered that yeast in the nectar led to both a reduction in the amino acid concentration and a change in the relative amounts of those amino acids, but rather surprisingly it did not affect the nectar sugars. Conversely, the bacteria increased the concentration of amino acids and the proportion of nectar sugars comprised of fructose and glucose, compared to sucrose, and lowered the volume of nectar. The scientists go on to speculate that these differing effects probably influence the interactions between the plant and its hummingbird pollinators.

It is really only in the last decade or so that we've learned just how complex the nectar ecosystem can be and how microbes can have dramatic effects on the ecology and evolution of flower–pollinator relationships. For example, the red tubular flowers of *Moussonia deppeana*, a member of the African violet family (Gesneriaceae), remain open longer and produce more nectar when they are infected by a fungus that is transferred by their hummingbird pollinators. Just think about that for a moment: here's a member of one biological kingdom, Fungi, manipulating the behaviour of species from two other kingdoms, Animalia and Plantae, in a complex ecological *ménage à trois* that suits its own evolutionary purpose. It's a remarkable finding, though I suspect that we are only just scraping the surface of how complex these relationships can be, because as well as yeasts and bacteria, flowers may contain a diversity of other kinds of organisms.

Nasal commuting

Living in Oxford in the 1980s and 1990s was a wonderful time for anyone like me who loves music and books. A cohort of great bands emerged from the 'Oxford Scene', including Radiohead, Supergrass, Ride and Swervedriver. But I was, and am, also a bookworm who enjoys spending an hour or two in a bookshop. At that time, pre-internet, Oxford was blessed with an abundance of second-hand and antiquarian booksellers, in addition to an Oxfam bookshop to which academics often donated their unwanted review copies and other freebie books. One of my prize purchases from that time, which I still own, is the book *Community Ecology*, a selection of chapters by different authors and edited by Jared Diamond and Ted Case. Unlike most academic texts of this vintage, this is a book I have referred to time and again throughout my career. It was also my introduction to a then little-known (at least to me) but intriguing aspect of the natural history of flower-visiting birds: hummingbird flower mites.

In the chapter by American ecologist Robert Colwell called 'Community biology and sexual selection: lessons from hummingbird flower mites', I learned about these intriguing players within the bird–flower world. Adult flower mites, which are arachnids distantly related to spiders and scorpions, are tiny – typically less than 1 millimetre in length. The females lay eggs within flowers and they, together with the immature mites, feed on nectar and sometimes pollen. At least 100 hummingbird-pollinated plants have been identified that contain these mites, though this is certainly an underestimate because many plants have not been investigated. Depending upon the type of plant, inflorescences can contain anywhere between five and more than 1,000 mites, and they can be very abundant within individual flowers. A recent survey of 14 plant species at La Selva Biological Station in Costa Rica recorded 10,654 mites from 489 flowers, an average of 22 mites per flower. In some plant species, the mites can reduce the amount of nectar available in flowers by up to half, though there's no real evidence that the reproduction of the plants is affected.

Individual flowers and inflorescences have limited lifespans. The mites must therefore make decisions about when to move to a new source of nectar, which might also be a place to lay eggs. Within a plant they typically walk from flower to flower. But between plants, in tropical communities where the distribution of individual plants is often patchy and infrequent, how do they make the move? This is where the hummingbirds come in: the mites, in effect, hitchhike on the birds, crawling onto their bills as they probe for nectar. And because the mites can easily be displaced from the bills, they seek sanctuary in the birds' nares, the nostrils through which they breathe.

It's such a fascinating interaction! And reading that chapter for the first time introduced me to a new word, 'phoresis', referring to animals that hitchhike on other animals, another form of commensalism.

Since beginning his research on hummingbird flower mites in the 1970s, Robert Colwell amassed a huge collection of more than 7,000 microscope slides containing specimens of almost 50 identified mite species, plus many that are probably new to science. In 2019 Robert donated all these slides to the Biodiversity Research Collections at the University of Connecticut, where they will be held in perpetuity as a valuable and unique resource for future researchers interested in this three-way interaction between bird, arachnid and plant.

The relationship between hummingbirds and the mites that inhabit flowers, then travel in their nostrils between plants, sounds extreme and unusually specialised. But in one sense it's not surprising: mites are fairly ubiquitous invertebrates that have adopted a wide range of lifestyles during their long evolution. The broad group that these flower mites belong to is known to date to at least the Cretaceous, around 100 million years ago, based on specimens found in amber from this period. It seems likely that all birds have mites associated with them, and at least 2,500 species have been described, so perhaps the evolution of this flower–mite–bird relationship was almost inevitable. At some point in the past mites have presumably discovered that there's an advantage to spending part of their time living in flowers rather than just on the birds.

It's not just hummingbirds in the New World which interact with mites in this way, however, and a recently described interaction between mites, fungi, flowers and the Cape Sugarbird shows just how intricate these relationships can be. In 2018, writing in the journal *Microbial Ecology*, a team of South African scientists showed that Cape Sugarbirds feeding on protea inflorescences pick up mites on their bodies, mainly in the feathers under and around the beak. It's estimated that a single bird can carry as many as 1,000 of these mites. But the story doesn't end there. Attached to those mites are smaller mites of a different species, which in a sense are hitchhiking on the hitchhikers – they are 'hyper-phoretic'. And these more diminutive mites, in turn, carry the spores of a fungus which grows within the protea inflorescences and which the mites, it appears, cultivate and feed on. The Orange-breasted Sunbird also carries the same larger mites, and these mites in turn are found in a range of protea species, many of which are bird-pollinated. This elaborate set of relationships is probably more ecologically widespread, and more complex, than we currently realise.

Honeyeaters and sunbirds in Australasia and Malaysia are also known to carry flower mites, but the details of these interactions are not as well worked out as they are in the other parts of the world that I've described. This is unfortunate, as one of the few groups of plants in which this has been noted are bird-pollinated mangroves, a key type of tree in our efforts to reduce the impacts of climate change in tropical regions.

Some spiders really do eat birds

Having said that nature is not all 'red in tooth and claw', we should be aware that sometimes it is, and just as all flower-visiting birds eat arthropods, sometimes those invertebrates turn the tables.

The term 'bird-eating spider' generally refers to the Goliath Tarantula of South America, the world's most massive spider. This arachnid can have a body length of up to 13 centimetres and weigh almost 10 times as much as the Giant Hummingbird, the largest in the family. The spider's bird-eating habit is, apparently, rather rare and derives from an early-eighteenth-century illustration in the book *Metamorphosis Insectorum Surinamensium* by the influential German naturalist and artist Maria Sibylla Merian (1647–1717). Her drawing shows the spider on a branch, atop a prone hummingbird adjacent to a small, cup-shaped nest containing four eggs. If predation of hummingbirds by this active ambush predator is rare, four eggs in a hummingbird nest is rarer still: two eggs per tiny nest is much more typical. Nonetheless, this illustration was almost certainly based on a real event, as Merian spent two years travelling, observing and sketching in Suriname and was a careful witness to the natural world. The English writer and explorer Redmond O'Hanlon suggests that Merian's observations were dismissed as a fantasy because she was a woman, and only accepted more than a century and a half later when Henry Walter Bates (1825–1892) described the spider killing two finches in his book *The Naturalist on the River Amazons*.

The Goliath Tarantula is an opportunistic ambush predator, so most of its diet consists of large insects, small reptiles and rodents, and other terrestrial animals. There's a recent published account of this spider consuming a Common Scale-backed Antbird that was captured in a mist net, but otherwise birds must feature infrequently in the diet of a largely ground-based arachnid. This is supported by an experimental study by Brazilian researchers João Vitor de Alcantara Viana, Murilo Massufaro Giffu and Leandro Hachuy-Filho, who studied the behaviour of hummingbirds visiting heliconia flowers on which large, dead tarantulas had been placed. The team discovered that the birds did not try to avoid those flowers, probably because they rarely encountered the spiders in nature and had not learned how dangerous they could be. The bird observations were backed up by data on the amount of pollen removed from anthers and the volume of nectar remaining in the flowers: both were indistinguishable from flowers of the control, spider-free heliconias.

Predation of hummingbirds by other types of spiders that create large webs does, perhaps not surprisingly, occur more often. In a survey of ornithologists, the biologist Dan Brooks unearthed accounts of 69 examples of birds that had become trapped in spider webs. These birds belonged to 54 species across 23 families, with hummingbirds accounting for almost one-third of the records and 16% of the species. Given that hummers are amongst the smallest of all birds, this is perhaps to be expected, though there are also records of much larger birds such as honeyeaters, sunbirds and even a Laughing Dove (*c.* 80 grams) becoming trapped. A few of the birds were found dead and wrapped in silk, having presumably succumbed to exhaustion and the spiders' venom. The spiders themselves were almost all large orb-weavers, the webs of which can span a couple of metres and are a discomforting encounter for any unwary ecologist, as I know from experience. For a small bird it must be truly terrifying to become trapped in these webs, though sometimes they can free themselves.

Dan's study was published in 2012, and since then there have been many more accounts of birds caught by spiders in webs, with hummingbirds again featuring prominently. A recent survey of Asian records found that white-eyes, sunbirds and other avian flower visitors are sometimes also trapped and consumed. Predation of these birds by spiders is not directly linked to their flower-feeding behaviour, though, unlike the predation of bees and flies by crab spiders, which sit in wait on flowers. It's the relatively large size of the spiders in comparison to the birds, plus the strength of their webs, that are the main factors. However, a review of bird predation by praying mantises led by zoologist Martin Nyffeler (whom we'll encounter again in Chapter 15) certainly did demonstrate a link with nectar drinking. This study concluded that more than 70% of the reports of mantises eating birds were from gardens in the USA, where they capture hummingbirds that are visiting feeders or nectar-rich flowers. As with the spiders, there are also occasional reports of sunbirds and honeyeaters being consumed by these insects.

Spiders and praying mantises are not the only invertebrates that occasionally feed on hummingbirds. There are written accounts of dragonflies doing the same thing, and wasps will sometimes attack them too. It must be tough being such a small bird. Yet, as we've seen, often the birds that spiders and other invertebrates consume are just part of a larger array of food items that make up the diets of most predators. The South American Bushmaster with which I almost had a too-close encounter in Guyana feeds on a range of rodents as well as reptiles, frogs, the occasional small monkey and, yes, birds. It has a wide geographic range across the continent and is a dietary generalist. Some of the species that it feeds on are undoubtedly more specialised in their food requirements, however, and that's the next topic we're going to turn to: just how specialised are flower-visiting birds and bird-pollinated flowers?

CHAPTER 11

The limits to specialisation

Nature documentaries, glossy coffee-table volumes and biology textbooks are full of examples of highly specialised interactions between birds and flowers. Often hummingbirds are the poster children of this narrative, with enthralling stories of how the dramatically bowed beaks of species such as sicklebills precisely fit the mirror-image curving corollas of certain flowers, especially heliconias.

Such relationships between flowers and birds certainly do exist, and they are forceful reminders of the power of evolution to sculpt precise interactions between species. But nature, as always, is more complex than these co-evolutionary stories would suggest; there's almost always more to the tale than meets the eye. Given what I said in Chapter 2 about hovering versus perching, it's significant that hummingbirds such as these often do not hover when foraging from heliconia flowers. Instead they land on or hang from the bracts, and drink the nectar standing still. This strongly suggests that the presence of one type of specialisation (a curved bill) limits the ability to engage in another type of specialisation (hovering to feed at flowers).

Not only that, but to focus just on the extremes of specialisation is, I feel, an injustice to the ecological and evolutionary realities of how these birds make a living and how the flowers, in turn, exploit the birds to their own advantage. People like dramatic accounts of co-evolution. But in the natural world it's often the mundane aspects that are the most important, rather than the specialised ways in which birds and flowers have co-evolved or come together. These more generalised interactions have been a recurring theme throughout the book, but in this chapter I want to present some of the more extreme specialist examples, with the caveat that things may not be as straightforward as we used to believe. Although these are the stories that are often highlighted, they are not the whole story, nor even representative of hummingbird–flower relationships, which many consider to be the most specialised of all.

Hummingbird expert Gary Stiles showed this complexity in the 1970s when he examined the relationships between the nine species of hummingbirds that visited the flowers of nine species of heliconia at La Selva in Costa Rica. Only one hummingbird, the White-tipped Sicklebill, possessed what Gary described as a 'very strongly decurved' bill, another was 'moderately decurved', and the remaining seven were 'slightly decurved' or 'straight'. Likewise, only one of the heliconias had a flower that could be described as having a 'fairly strong' curvature, the others being straight or slightly to moderately curved. As per the textbooks, the only heliconia that the White-tipped Sicklebill visited was that species with the strongly curved flowers. However, and this is not usually highlighted, the flowers of that species were also foraged on by all the other hummingbirds. From the birds' perspective, this was not a specialist resource only accessible to species with an extremely curved bill. Not only that, but birds other than the sicklebill could pick up pollen on their heads and bodies, suggesting that this heliconia is pollinated by a range of hummers with different bill shapes and behaviours, including species that defend a defined territory and those that trapline and fly a set circuit among patches rewarding flowers. The fit between the flower and the bill in this pair of species may, as Gary suggested, reflect specialisation of the plant on the White-tipped Sicklebill in line with Stebbins' most effective pollinator principle. Yet, as I discussed in Chapter 5, it could also be an example of Aigner's least effective pollinator principle, in that the curved flowers do not exclude other bird species from pollinating them.

There are other examples of close hummingbird–flower relationships that are more than meets the eye, once that eye has taken a wider and more detailed view of the relationship. On the Lesser Antillean islands of St Lucia and Dominica, the flowers of two heliconia species were suggested by Ethan Temeles and co-workers to be only pollinated by, and tightly co-adapted to, one hummingbird species, the Purple-throated Carib. Later work by Bo Dalsgaard and others showed that the story is actually more involved than that. They suggest that the interactions between heliconias and hummingbirds in the Lesser Antilles may actually be a 'geographic mosaic', a phrase coined by American biologist John Thompson to refer to the way in which locally co-adapted, specialised interactions form part of a more widespread set of looser relationships. In this case, other hummingbirds such as the Green-throated Carib, Rufous-breasted Hermit, Blue-headed Hummingbird and Antillean Crested Hummingbird are thought to also play a role in heliconia pollination on Dominica and Grenada.

In addition to the curved beak and flower examples, stories are told about species that are engaged in interaction 'arms races' as bird and flower systematically evolve longer and longer bills and corollas in unison, to the exclusion of other species. These also deserve discussion, but before we go any further I think it's important that we ask a fundamental question.

What do we mean by 'specialised'?

Sicklebill hummingbirds are not the only avian flower visitors with elaborately curved beaks. On the other side of the Atlantic they are almost rivalled by some African species such as the large Golden-winged and Sao Tome Sunbirds. It's the Hawaiian honeycreepers such as the Iiwi and the sadly extinct Black Mamo that have the most elaborately curved bills of all, however. Their beaks arc down through almost 90 degrees and, as expected, there are flowers on Hawaii which match. All of this speaks to 'specialised' relationships between flowers and birds. But ideas about specialisation mean different things to different people. That's partly why I chose 'limits to specialisation' as the title for this chapter: it's not just about the limitations on how specialised plant–pollinator interactions can be, it's also about the limits to how different scientists and naturalists understand that word and those interactions.

To try to bring some clarity to this area of pollination ecology, in the past I've written about different forms of specialisation, using qualifiers such as ecological, functional and phenotypic. Ecological specialisation refers to the number of partners a species has: a flower that is pollinated by just one species of bird is a strict specialist, whereas a flower that is serviced by many different types of birds is less specialised. Both of these flowers would be classed as functional specialists, though, in the sense that they both use a single functional group of pollinators – birds.[1] Phenotypic specialisation is a slightly trickier concept, but, in essence, for flowers it relates to how closely adapted they are to their pollinators, in comparison to their closest evolutionary relatives.

We can also look at all these types of specialisations from a bird's point of view. Hummingbirds have evolved a series of phenotypic specialisations that allow them to feed on nectar, including relatively long, slender bills and highly extensible tongues that act as fluid pumps, the latter a quite recent discovery by Alejandro Rico-Guevara, Tai-Hsi Fan and Margaret Rubega. Other birds that specialise on nectar for at least part of their diets, for example sunbirds and honeyeaters, have tongues that are highly frayed and function more like mops, soaking up the fluid.[2] Similar tongue morphologies have also evolved independently in the warblers, such as the Fire-tailed Myzornis mentioned in Chapter 3. Other adaptations to nectar feeding include aspects of the birds' physiologies and digestive systems, including enzymes that allow them to break down nectar sugars more effectively.

The possession of specialised nectar-feeding anatomy such as highly adapted tongues does not necessarily mean that a bird is a more effective pollinator; it simply means that it can exploit a major food source more successfully. Some types of flower have in turn adapted to the birds' abilities, and their colour, shape, size and so on reflect that adaptation. But not always: birds can

---

1 Some would argue that, say, long-billed and short-billed hummingbirds belong to different functional groups, or hummingbirds and passerines are likewise functionally different. But this simpler way of looking at it will suffice for now.

2 I've never understood why these are often referred to as 'brush-tipped' tongues. Brushes brush, they don't soak up fluids. I suspect that whoever first named them had never encountered a mop.

effectively pollinate flowers that possess no bird-adapted characteristics at all.

Beyond ornithophily

During the month that I spent at the Mpala Research Centre in Kenya, teaching on a Tropical Biology Association field course, I frequently encountered Variable and Mariqua Sunbirds feeding on flowers within the savanna ecosystem (Plate 27). All of the plants from which they took nectar, including various acacia species and a type of milkweed growing in rocks in the river, had flowers which in no way resembled the archetypical 'ornithophilous' flowers. The acacia flower heads were not brightly coloured tubular structures but white and cream pompoms, and the main pollinators were the many insects – flies, bees and butterflies – that visited them in great abundance. Likewise, the milkweed flowers were white and open, and in any case the pollen-packaging strategy that they use means that sunbirds are unlikely, though not impossible, pollinators (see Chapter 6).

Kenya during this period was experiencing a severe drought, and everything was looking very dry. These flowers were the few available to birds that require a significant proportion of dissolved sugars in their diet. No doubt when the rains finally arrived so too would flowers that were well adapted to bird pollination. We can think of these specialised bird-pollinated flowers as being embedded within a background noise of generalist flowers on which the birds also depend. As far as survival of the birds is concerned, they don't really care what a flower looks like as long as it supplies a reliable flow of nectar. And there's an argument to be made that the specialists are less important than the generalist flowers to bird survival, if it's the generalists that are getting the birds through periods of drought.

On the other side of the world, Pietro Maruyama has studied this question in a very different type of savanna within the Brazilian Cerrado. Pietro and his team spent a few seasons assessing the hummingbird–flower interactions in two sites, one from 1996 to 1997, the other from 2009 to 2011. The flowers of the

24 species of plants observed across the two sites were categorised as ornithophilous (i.e. showing specific adaptations to bird pollination) and non-ornithophilous (with no such adaptations). In the first site, hummingbirds visited 18 plant species and 10 of them (56%) had typically ornithophilous flowers. In the second site, of the nine flowers which had hummingbirds as visitors, four (44%) were ornithophilous. Not only that, but the researchers calculated that the non-ornithophilous flowers often contributed as much, and sometimes more, energy to the birds, especially at times of the year when there were few ornithophilous flowers available.

One such example is a species of leguminous tree in the genus *Bauhinia* that has flowers which open at night and are mainly bat-pollinated. The team worked out that this one species alone accounted for a five-fold increase in the availability of energy across the landscape for the birds, despite the fact that they probably contributed little to pollination (though that remains to be tested).

The impact on flowers of birds as nectar thieves is not likely to be great, at least for mass-flowering plants such as acacias that produce an enormous number of flowers, only a fraction of which are ever going to produce seeds. The number of birds is usually low in comparison to the insects, which collectively take far more nectar (and pollen in the case of bees and some flies). So anti-bird defences are rare in nature, as far as we know. On the other hand, anti-bee defences, in which bird-pollinated flowers have evolved to protect themselves from the depredations of bees, do seem to be quite common. As discussed in Chapter 8, these defences include a lack of odour, and a flower colour to which bees are not especially sensitive. They may also include flower traits that affect how pollen is transferred, such as the extent to which the reproductive parts of the flower project beyond its opening, and the angle at which the flower is presented. In addition, bitter-tasting compounds might also deter bees, as we saw in Chapter 6.

It is not only hummingbirds in the New World that show this pattern of visiting a wide range of different flower types. So too do sunbirds in Africa, as documented by Kryštof Chmel and colleagues. In a study aptly titled 'Bird pollination syndrome is the plant's adaptation to ornithophily, but nectarivorous birds are not so selective', these authors concluded that

> ornithophilous plants were not more attractive than the other plants for nectar-feeding birds. *Nectar production per plant individual* was a better predictor of bird visitation than any other floral trait traditionally related to the bird pollination syndrome.

I have emphasised some words there, because this is a key message: birds do it for themselves, not for the flowers. It shouldn't surprise us that nectar-feeding birds will investigate, and exploit, any flowers with the potential to provide them with sustenance. After all, hummingbirds and sunbirds regularly visit artificial sugar-water feeders which are in no sense 'natural'. The value of such knowledge is that it broadens our understanding of the ecology, and potentially the conservation, of these birds by acknowledging that they are more than just actors in co-evolutionary dramas. As important as such specialised pollination relationships are, especially for driving plant evolution, they are only part of a bigger jigsaw puzzle. We need to focus on the wider picture, on the network of interactions and physiology and behaviour that sustain these birds and their contributions to ecosystem functioning in many regions of the world.

What does it mean for the flowers?

At this point, sceptical readers, brought up on documentaries of close bird–flower adaptations, might shrug their collective shoulders and say: 'So what? Just because birds can take nectar from a wide range of flowers with different features, that doesn't mean they pollinate those flowers.' That's a fair point, and until recently the assumption has been that if a flower looks like it *should* be bird-pollinated then it *is*, but if it *doesn't* it's *not*.

There's a circularity to that argument, however, which means that the effectiveness of birds as pollinators is usually only tested on flowers that, *a priori*, are assumed to be bird-pollinated because they fit the classic ornithophilous model. On the other hand, experiments which test the effectiveness of birds as pollinators of non-ornithophilous flowers, such as that performed by Stella Watts on *Duranta* (see Chapter 7), are relatively rare. In that example, of course, we found that birds do not effectively pollinate the flowers. This was in line with the floral colour and morphology – but we can't assume that this is always going to be true.

In fact, we *know* it isn't always true. Back to Brazil, and more work by Pietro Maruyama and André Rech, this time led by Marsal Amorim. Working in the rupestrian grasslands of Minas Gerais, they found that hummingbirds effectively pollinated a range of different types of flowers, from classically ornithophilous through to flowers having none of the typical traits for bird pollination. To quote the authors, their 'results indicate that hummingbirds contributed to plant reproduction independently of the fit to bird pollination syndrome'.

It's taken researchers of bird–flower interactions a long time to really appreciate this point. While he was reading a draft of this chapter, Nick Waser told me that in the 1970s:

> When I tried to get a doctoral dissertation grant of a few thousand dollars from the US National Science Foundation … my application was quickly declined because some of the species I studied for my PhD did not fit the syndrome – most notably the Blue Larkspur … As I showed then, and Mary [Price] and I explored further, this flower is about equally pollinated by hummingbirds and queen bumble bees at the Rocky Mountain Biological Laboratory.

It's possible to trade examples like this back and forth, in which floral traits sometimes predict bird pollination and at other times do not. The important thing that I want to highlight is something that I said earlier in the book: it's often possible to predict that a flower *is* bird pollinated, but not that it *isn't*.

Just to emphasise how our understanding of the complexity of the role of birds in plant reproduction is changing, consider some recent work by Felipe Amorim. In the Atlantic Forest of southeastern Brazil, Felipe and his team encountered the bizarre flowers of a plant called *Scybalium fungiforme*, which as the species name suggests is a secretive and rather fungus-like parasite of other plants. They predicted that these flowers would be pollinated by ground-dwelling mammals, and using infrared cameras they showed that opossums are important visitors and potential pollinators. Shortly afterwards, however, they had to revise their views when they discovered that bats and birds also play a role in pollinating this weird plant. Ruby-crowned Tanagers probably don't pollinate the flowers themselves, but they do remove the bracts that cover the flowers, in order to feed on the nectar (Plate 12). In fact they are the *only* birds of those so far observed that are able to do so, and this in turn allows other visitors to access that nectar. These are as varied as wasps, bees, several rodent species (even a squirrel), coatis, Tayras, bats, hummingbirds and a dozen other bird species. Some of the latter, Felipe tells me, include species previously not known as nectarivorous. The effectiveness of these visitors as pollinators needs to be tested, but Felipe's preliminary description is that the tanagers are 'facilitators' of other pollinators in these flowers. It's fascinating, ongoing work and I look forward to seeing how it develops.

How specialised are the birds?

When it comes to ecological and functional specialisation, from the bird's perspective things are even more complicated. There are no phenotypically specialist birds that feed on nectar from the flowers of just one species of plant, though at any one time or place it may appear that they are this specialised because there are no other suitable flowers for them to visit. Less specialised birds, such as warblers and tits, may only ever consume the nectar of one type of plant, though again as generalists they take whatever is available to them.

We can dispel the notion that there are large numbers of one-to-one flower–bird relationships that are so tightly linked that they exclude other species. The one documented example that I thought might exist is the interaction between Least Seedsnipes and a slipper flower that I discussed in Chapter 6. However, the lead researchers on this work, Alicia Sérsic and Andrea Cocucci, recently told me that they now know that seven other bird species of the families Passerellidae, Thraupidae and Mimidae visit this slipperflower and probably act as pollinators, so this may not be such a straightforward mutual-exclusivity story. Not only that, but natural hybrids exist between the seedsnipe-pollinated species and a second, bee-pollinated slipper flower, implying pollen movement between these very different pollination systems. Just to confound matters further, the Least Seedsnipe has been documented feeding on nectar from the flowers of a legume in the Patagonian steppe, though the most frequent pollinator of that plant is the Rufous-collared Sparrow.

Perhaps we shouldn't be surprised that interactions turn out to be less mutually specialised than first thought. One-to-one plant–pollinator interactions are extraordinarily rare in nature, and most examples have either evolved from seed-predation relationships (such as the pollination systems in which figs depend on fig wasps and yuccas on yucca moths) or involve deception of insects that are not normally flower visitors, such as flies looking for food or rotting organic material on which to lay their eggs, or orchids that mimic female insects and lure males into attempting to mate with them.

As I've already noted, specialist as opposed to generalist pollinating birds are typically characterised by beaks and tongues that are well adapted to consuming nectar from flowers. Specialised beaks are often narrow and pointed, ideal for probing within the deepest of flowers, and the tongues may be highly fringed, to act as a mop, or partially tubular, to pump up the bird's fluid diet. Add in aspects of their physiologies such as specialised enzymes for breaking down sugars, and we can be quite certain that such adaptations point to birds that have evolved to take full advantage of the sweet bounty offered by flowers.

But again, we need to consider what this means for those blooms. Does it indicate that these specialist nectarivores will always be the best pollinators of the flowers that they service?

Often the answer is yes, and there are lots of examples of plants that have evolved to be pollinated by just one type of hummingbird or sunbird, or by all of the available birds, but birds alone (insects not allowed). We can't take this for granted, however, and sometimes birds and flowers defy expectations, as Carolina Diller, Miguel Castañeda-Zárate and Steve Johnson found when they investigated the pollination ecology of one of the tree aloes in South Africa. When these tall, succulent plants (relatives of that mainstay of health and beauty products, *Aloe vera*) bloom they produce long inflorescences containing almost 300 bright orange tubular flowers that are full of nectar. These flowers are visited by many generalist birds including Black-capped Bulbuls, various weavers and Cape Glossy Starlings, as well as specialists such as Amethyst Sunbirds. Using single-visit deposition (SVD) experiments of the type I described in Chapter 7, the team assessed the relative importance of these two groups of birds, the specialists and the generalists. The results challenge the idea that it's always the specialists that are the best pollinators: those sunbirds deposited no additional pollen on stigmas compared to flowers that had been bagged as experimental controls. The generalists on the other hand deposited about three times more.

In one respect, however, Carolina and the team were not surprised at their findings. They had predicted that the opportunists would be the main pollinators because the nectar of this aloe is hexose-dominated, consistent with a plant that has evolved to attract and reward generalist rather than specialist nectarivores (see Chapter 6). Consequently, more generalists visited the flowers than sunbirds, but that doesn't explain why the sunbirds are not better pollinators of these plants. On average the beaks of these two groups of birds are about the same length, but the generalists have thicker beaks and larger heads, which the researchers think are the key factors. As they put it:

> Sunbirds with their slender bills, small heads and longer tongues seldom make effective contact with the reproductive parts of the flowers.

There are times, however, when having a slender bill makes for a very effective pollinator, and for other South African plants, Sjirk Geerts and Anton Pauw have described 'hyper-specialization for long-billed bird pollination' in a set of plants within the Cape Floral Region that rely on Malachite Sunbirds as pollinators (see Plate 2). To truly appreciate the extent to which a long-billed bird can drive the evolution of deep flowers, however, we need to skip back to South America.

The sword bearer

I can clearly remember the first time that I encountered the Sword-billed Hummingbird. Silhouetted against the late afternoon sky, the bird paused on its shrubby perch and slowly looked around. It wasn't until its head was sideways on to me that I realised what it was, a specialist of the high Andes whose scientific name – *Ensifera* – means 'sword-bearing'. The beak was unmistakable, at 10–11 centimetres it's the longest of any hummingbird and seems unnecessarily exaggerated as a structure for feeding from flowers. Over the subsequent days we saw it occasionally visit some of the passionflowers with equally long floral tubes, well adapted to a bird of its dimensions.

Indeed, the Sword-billed Hummingbird is often cited as a classic example of co-evolution with flowers, and work led by Stefan Abrahamczyk suggests that over the last 10 or 11 million years almost 40 species of Andean passionflower have evolved in response to the natural selection associated with pollination by such a long-billed bird. However, this evolution is not one-way, and passionflowers pollinated by bees, bats and short-billed hummingbirds have all evolved from Sword-billed-pollinated species. This challenges some earlier suggestions that once a species has evolved that type of pollination system, it's irreversible and can't evolve relationships with other types of pollinators. None of this, though, answers a more fundamental question: why does the Sword-billed Hummingbird have such an extremely long beak?

This species is one of the most spectacular and impressive of all flower visitors, and it's certainly an icon of South American ornithology and pollination biology. The extreme beak length by which it gained its name renders it unique among birds: no other living species has a bill that exceeds its body length. Despite this extreme specialisation in bill morphology, its diet, like that of all hummingbirds, also includes insects. The sword-bearer is a widespread and relatively common species in the high Andes – though, surprisingly, its nest was not described until 2022, when William Arteaga-Chávez and colleagues documented some fundamental aspects of its reproductive biology.

The usual explanation for the impressive bill is that the birds have co-evolved with their main nectar sources, especially passionflowers and daturas, in one of those arms races that I mentioned previously. Put simply, long-tubed flowers benefit from the birds having to probe more deeply, so picking up or depositing more pollen. This drives the elongation of the flowers, which in turn means that the birds must evolve ever-longer beaks to access the nectar. There are some problems with this story, however. The implication from a lot of the published literature on the Sword-billed Hummingbird is that it only visits flowers with long corolla tubes. That's far from the case – these birds will feed on a range of flowers, not just the long-tubed species, as well as artificial nectar feeders that lack any kind of tube at all. Evolving such a long beak imposes a cost on these birds in terms of resources and energy. If they can achieve their nectar requirements from short-tubed flowers, which are by far the most common type in the habitats in which the birds live, why should they engage in this evolutionary arms race?

There is an alternative explanation that I think deserves to be investigated. Extreme morphologies in birds are often the result of sexual selection, or at least male–male competition. Individuals with the most extravagant plumage (think peacocks – male Indian Peafowl), the most mellifluous songs, the best dance moves (birds of paradise) or building abilities (bowerbirds) have more frequent opportunities to mate.

There's certainly evidence of sexual selection going on for the beaks of other hummingbirds. Alejandro Rico-Guevara and Marcelo Araya-Salas have suggested that some hummingbird bills have evolved as weapons. When they tested their 'bills as daggers' hypothesis, they found that male birds with bill tips that are larger and more pointed had greater success defending their territories within leks.[3] When male hummingbirds fight, as they often do, they stab rivals with the tip of the bill.

As far as I know, however, it has not previously been suggested that the extreme beak length of the Sword-billed Hummingbird might be due to sexual selection. An obvious issue with this idea is that it's both sexes that have the long beaks, not just the males, and in fact the male has a slightly shorter beak on average. There are, however, precedents for mutual male–female sexual selection in the bird world, and indeed it was a subject about which Charles Darwin speculated. The first experimental test of mutual sexual selection was not until the early 1990s, however, when Ian Jones and Fiona Hunter showed that both male and female Crested Auklets prefer members of the opposite sex that have longer head crests. Since then, evidence has accumulated that males may also be choosing females with particular ornamental traits in other species, for example the Bluethroats studied in Norway by Trond Amundsen's research group. Whether these preferences are reciprocated by the females is yet to be determined, but the findings are certainly suggestive.

In the scenario I'm suggesting, the elongated floral tubes of passionflowers, daturas and other plants have evolved as a consequence of the long bills of the Sword-billed Hummingbird. Those flowers that most effectively place pollen on and remove it from birds are selected for, and have likely evolved as the birds themselves have evolved their elongated mouthparts. But that's a result of male–female sexual selection in the birds, not an arms race between birds and flowers. Of course, this is highly speculative and requires further, preferably experimental, study.[4] Nonetheless it's a good example of one of the main themes of this book: sometimes

3 A lek is an area used in the breeding season by male birds that engage in communal displays in order to attract a mate.

4 To prove to myself that this idea was not completely deluded I sent this section to Alejandro and was pleased to find that he did not dismiss it. In fact he replied with some hypotheses that I hope he'll test in the future.

the simple textbook explanations of the natural world are not all that they might appear.

Before we leave this topic we should remind ourselves that that there is no binary distinction between 'specialist' and 'generalist' birds or flowers. Depending on how we define and approach specialisation, there's clearly a continuum from the most specialised to the most generalised species, along which all flowers, and all their pollinators, can be located. As nectar feeders, hummingbirds and other avian flower visitors can be extreme generalists, taking nectar from whatever sources are available. As pollinators, however, they sometimes leave an indelible evolutionary imprint on the flowers that they service. In Chapter 4, for instance, we saw that interactions with hummingbirds seem to have speeded up the diversification of the South American gesneriads over the last 25 million years or so, since these birds and flowers first encountered one another in the Early Miocene. Much of this evolution has taken place in the mountains of the Andes, and it's isolated mountains and oceanic islands which have provided us with some fundamental insights into the ecology and evolution of bird–flower relationships.

CHAPTER 12

Islands in the sea, islands in the sky

In silent awe, my students and I watched as the curator donned his white cotton gloves, unlocked the case, and slid out the tray of bird specimens. For several years I had taken a group of first-year undergraduates for a behind-the-scenes tour of the Natural History Museum's ornithological collection at Tring in Hertfordshire. This, however, was the first time that we'd been offered the chance of seeing the original specimens of 'finches' collected by Darwin in the Galápagos Islands. Each of the preserved, stuffed birds was stiff and formal, head, body and legs aligned along an unnatural axis. We listened as the curator explained the significance of the birds and showed us the only remaining specimen label written by Darwin himself. After some polite questions from the students, the birds were returned to their secure cabinet, arguably the world's most valuable ornithological specimens, culturally as well as financially.

The finches of the tropical Galápagos Islands are icons of the bird world, a 'must tick' on the life lists of many birders,[1] and one of the most studied groups of species in the world. Their scientific discovery by Charles Darwin during his *Beagle* voyage in the early 1830s is a frequently recounted story in evolutionary biology. From an initial, very small, founding population the finches have diversified into about 18 species that exploit a wide range of habitats

1 Though that's not without its problems – the islands are under increasing pressure from unsustainable 'eco-tourism' that threatens to destroy the very things that people come to see.

and feeding niches, from seed eating to blood sucking.[2] They also visit flowers. Before we explore the relationship between Darwin's finches and flowers, however, I'd like to take a little detour into the story behind the discovery of how important these finches were for understanding evolutionary patterns, including some personal history.

Darwin did not fully recognise the significance of the finches when he first encountered them, and it was not until 1845, and the second edition of his book about the voyage, that he delivered the famous line:

> one might really fancy that from an original
> paucity of birds in this archipelago, one species
> had been taken and modified for different ends.

It was another British biologist, David Lack, who brought the finches to wider attention when he visited the Galápagos Islands in the 1930s and subsequently wrote up his findings in the 1940s. Lack was encouraged to follow in Darwin's footsteps and visit these remote islands by Sir Julian Huxley, the grandson of the great Thomas Huxley, Darwin's confidant and 'bulldog', and one of the most brilliant biologists of the nineteenth century. When David Lack returned to Britain, and following war service, he established himself at Oxford University as Director of the Edward Grey Institute of Field Ornithology. He also raised a family – and his son Andrew, whom you met at the beginning of the book, was one of my PhD supervisors. It's usual in the British doctoral system to have two supervisors, and my second was the late Denis Owen, who had been one of David Lack's research assistants at the Edward Grey Institute in the 1950s.

With an intellectual pedigree of ornithology stretching all the way back to Darwin himself, you might imagine that I have a long passion for birds. To my shame, that's not the case. As a youngster I had a moderate interest in birds, but as I grew older it was

---

2 'About' 18 species because some species are clearly hybrids, species boundaries are in some cases fuzzy, and speciation is actively going on in these birds.

exotic flowers, succulent plants and insects that really fired me up. Even as I got to know Andrew during my doctoral studies, I still considered birds a bit commonplace. Everyone was interested in them, and therefore I was going to choose the path where no one goes. As I described in the Introduction, it was Andrew's gift of a field guide to Australian birds after I completed my PhD that really ignited the slow burn of my avian interests. I still have the book, and its tatty pages have occasional scribblings of date and place next to birds that I successfully identified. Many I did not. It was not until I began to collaborate with serious ornithologists that I saw the importance of birds as pollinators, and how useful they are as a study system. Research careers often take a convoluted path, and most scientists will tell you that they got to where they are by following their noses rather than any linear, pre-planned trajectory.

But let's head back to the Galápagos. As I noted in Chapter 6, two of Darwin's finches, the Common Cactus-Finch and Medium Ground-Finch, have long been known as flower visitors that feed pollen to their young. Despite this, much of the interest in how those birds evolved into different feeding niches has focused on the importance of seeds, with larger, heaver beaks being capable of cracking open tougher seeds. More recently, Spanish ecologist Anna Traveset and colleagues undertook the most extensive study of bird–flower interactions ever attempted in an island setting. During four years of hard field work they showed that at least 19 of the 23 native Galápagos land birds (83% of the total species) feed on flowers and transport pollen on their bodies. This includes most of the Darwin's finches, plus the endemic Galápagos Dove and Galápagos Flycatcher, four species of mockingbird, and the Yellow Warbler. Their work also added a new family and order to Chapter 3's table of possible pollinating birds: the cuckoos (Cuculidae, Cuculiformes) in the shape of the Dark-billed Cuckoo.

More than 100 plant species are visited by these birds, the most frequent being two invasive species – the Common Guava, native to the Caribbean, and the Spotted Snapweed from India and Myanmar. Overall, though, more than 70% of the bird-visited plants were native to the Galápagos.

None of the birds could be described as nectar specialists, and indeed most were previously thought to be either herbivores subsisting on fruit and seeds, or specialists on insects, spiders and other arthropods. The broadening of the feeding habits of the Galápagos birds is attributed by Anna and her co-authors to a phenomenon known as 'interaction release'. This seems to be common in island situations (and perhaps elsewhere), and it occurs not just in birds but in other animals and plants too. In a nutshell, interaction release happens when species colonise islands, diversify into different forms, and begin to use resources that their mainland ancestors would have ignored or been excluded from because of competition from other organisms. For plants, one of the outcomes of interaction release is that they co-opt pollinators that are rarely or never used on the mainland. Mirroring this, animals that normally do not visit flowers begin to function as important pollinators. One of the most celebrated examples is lizard pollination, most instances of which are found on islands.

If this phenomenon seems at first rather insignificant, recall that there are tens of thousands of such islands scattered across the seas of the world, and that species move in both directions, including back to the mainland after having evolved in oceanic isolation. Interaction release on islands can thus act as a source of evolutionary novelty with global significance for biodiversity. To explore this further, let's leave the Galápagos and travel to the other side of the world, to an archipelago that Darwin wanted to visit but was prevented from setting foot upon: the Canary Islands.

The bird flowers of Tenerife

Good data are hard won. That's one of the main things that non-scientists, or undergraduates at the start of their scientific careers, often do not fully appreciate. Making observations, setting up experiments, collecting data is laborious, time-consuming and often quite boring. Without an eye on the prize (of scientific discovery and insightful revelations) it's easy to lose heart and give up. This is especially true of ecological field work – where, in addition to going about the business of data collection, one must often endure less-than-ideal living conditions. If he was initially naive about what he would have to endure, Darwin became well aware of this. During the three-year voyage of the *Beagle* he regularly complained about the food and lodging, not to mention sea sickness. But nonetheless he persisted: the drive to explore and discover overcame the discomfort.

Over the years I've thought a lot about Darwin's experiences when making annual trips to the island of Tenerife with students. I took to referring to it as 'Darwin's Unrequited Isle' due to the fact that the crew of the *Beagle* was prevented from disembarking because of concerns about a cholera outbreak back in Britain. It's tempting to speculate whether Darwin's evolutionary ideas might have begun to crystalise earlier if he'd been able to explore the Canary Islands. There are certainly examples of adaptive radiations of species across the archipelago to rival the Galápagos, especially of plants. There are also birds such as the Tenerife Blue Chaffinch which are found nowhere else in the world.[3]

Some of these Tenerife birds were suspected to act as pollinators, and in 2005 we tested this idea during field work led by Louise Cranmer in the mountains of the northeast of Tenerife. Louise had recently completed her PhD with my research group in Northampton, studying how pollinators use linear features such as hedgerows to navigate around the landscape. We'd applied for some funding with Lars Chittka for a postdoctoral research project to look in more detail at the 'bird flowers' of the Canary Islands. Louise and her partner Steve had rented a small property in the Anaga Mountains for a few months, and when I joined them for a couple of weeks it was not quite what I was expecting. My experience of Tenerife until then had been the dry, sunny south of the island. The other end of Tenerife was very different, the climate much colder and wetter. Inside the property the walls were wet, the bedding damp, and every wooden surface was coated with a fine bloom of greyish mould. In fact, we were living

3 The population of blue chaffinches on Gran Canaria has recently been determined to be a separate species, described as 'Europe's rarest songbird' because of its limited geographic range on that island.

in a cave. Or to be precise, half of the house was a cave, scooped back into the volcanic geology that forms the Anaga range. The front of the house had been built out from this, but nonetheless, the whole of the interior at this time of the year was cold and humid.

Tenerife has a surprisingly diverse climate, depending where you are on the island. The south and west is warm and dry, which is why the tourists flock there in their thousands every year; but the north and east, where the trade winds rise as they encounter the mountains, can be very wet and cold, even in the summer months. It's in this variable and at times unpredictable climate that some extraordinary flowers have evolved. I discussed these flowers at some length in *Pollinators & Pollination*, but I think the story bears repeating.

In the 1950s, the late pollination biologist Stefan Vogel visited the Canary Islands and wrote about a group of plants that he encountered which, in his opinion, possessed many of the traits typical of bird-pollinated flowers. These included characteristics that I've already mentioned in previous chapters, such as flowers that are rather sturdily constructed, with bright reddish or orange petals, and producing quite large volumes of nectar. Subsequent work showed that these traits were possessed by at least 16 plant species belonging to a diversity of families and genera, including some that are not typically associated with bird pollination such as the bird's-foot-trefoils (*Lotus*), buglosses (*Echium*) and foxgloves (*Digitalis*). They are mainly found in the cool, humid laurel forest that clothes the higher, wetter parts of the Canary Islands (Plate 14).

Stefan Vogel's original idea was that these plants were 'Tertiary relicts', the remnants of plant communities that were once much more widespread across Europe and North Africa over the past few tens of millions of years.[4] Specifically, he saw them as part of a flora that had covered the Sahara and the Canary Islands in the Pliocene

4 The 'Tertiary' period, which began with the demise of the non-avian dinosaurs 66 million years ago and ended 2.6 million years ago, is now known as the Paleogene and Neogene periods. The term 'Tertiary', however, survives as a relict in 'Tertiary relict'.

(5.3 to 2.6 million years ago). In this scenario, the 'bird flowers' had evolved to be pollinated by sunbirds, which had then become extinct as the climate became drier, reducing the vegetation to just a few patches on islands such as Tenerife and La Gomera. There is, however, no avian fossil evidence to support this idea, and I think that the argument is unconvincing, because if those pollinating birds had gone, why had the flowers that depend on them not also been rendered extinct?

The Canary Island bird flowers are visited by generalist species such as tits and warblers. Vogel's explanation was that these birds were functioning as 'secondary pollinators' that had learned to exploit the nectar bounty which the flowers provided long after the flower traits had evolved under the influence of the hypothetical sunbirds. As soon as I read about these flowers I was sceptical of Vogel's sunbird extinction idea. It seemed to me to be an example of circular reasoning: the flowers must be bird-pollinated but there are no specialist birds therefore the birds must be extinct. Vogel had a wealth of experience of working in the tropics and South Africa, and he knew that hummingbirds and sunbirds were associated with certain flower types, and had presumably seen them visit 'atypical' flowers, but it never occurred to him that non-specialist birds could also drive the evolution of flowers.

The story of the Canary Island bird flowers was well known. Several people – including Danish researchers Yoko Dupont and Jens Olesen, and Spanish scientist Alfredo Valido – had made observations of the visiting birds and the characteristics of the flowers, such as colour and nectar chemistry. But until Louise took a closer look and performed experiments on Tenerife, no one had actually demonstrated that these generalist birds were effective pollinators of the flowers. What Louise did, with help from her field assistants, was to measure the amount of pollen that the birds removed from anthers and deposited on stigmas when they visited flowers. It was detailed, painstaking work, in often trying conditions – as I said earlier, good data are hard won – but by the end of the field season she had collected enough data to show that birds such as the Canary Island Chiffchaff are actually excellent pollinators of at least three of these plant species. Working in Lars's laboratory back in the UK, Louise also showed that the flowers, when viewed from the perspective of bees and their ability to discriminate between colours, are likely to be poorly detectable by these insects. This matched the observations that were made at the same time by Ralph Stelzer, that the local bees tended to ignore the bird flowers.

One of the most interesting discoveries that we made was that the bird flowers on Tenerife have long lifespans, typically of 2–3 weeks, in keeping with reliance on pollinators that do not rely on nectar and only occasionally visit flowers. Pollination ecologists have known for some time that flower lifespan negatively correlates with rate of visitation by pollinators, such that frequently visited flowers are open for just a short period, often less than one day. During our time studying them, we found that on average the flowers of the Canary Island Bellflower (or Bicácaro) were visited only once every five and a half hours, those of the Canary Island Foxglove once every 16 hours, and a staggering once every seven days for an endemic Canarian bird's-foot-trefoil species. Furthermore, work by Yoko Dupont found that these Canarian bird flowers all have nectar that is hexose-dominated, with only small amounts of sucrose. As I described earlier, this is characteristic of flower feeding by generalist passerines rather than specialised nectar feeders such as sunbirds.

Later studies by other researchers showed that generalist birds effectively pollinate other members of the Canary Island bird flower alliance, such as an endemic and rather rare member of the mallow family (Malvaceae). However, some of the plants that were originally thought to be part of this group have been shown to be pollinated by insects and/or lizards, as well as birds.

All of this says to me that Vogel's hypothesis that these flowers were originally sunbird-pollinated is incorrect: when their ancestors arrived on the islands millions of years ago, the opportunity for interaction release meant that they evolved relationships with pollinators that their relatives on mainland Africa did not use. For some of these species we can date their arrival: for example, bird pollination in the endemic bird's-foot-trefoils seems to have developed about 1.7 million years ago. It's not just on the Canary Islands that this has occurred – Jens Olesen and colleagues have shown that adaptations to bird pollination evolved in several other species of bellflower on islands such as Réunion and Mauritius.

Stefan Vogel laid the foundations for a story of ecology and evolution on the Canary Islands which was more complex than he could have realised. As I've shown earlier in the book, pollination by generalist passerine birds, either alone or as part of a broader set of pollinators, including bees and specialist birds, is more common than previously thought. The fact that scientists have only recently appreciated this may be due to our preconceived ideas of what nectarivorous, flower-pollinating birds 'should' look like – specialists with bills, tongues and behaviours that are highly adapted to flower feeding. Flowers, on the other hand, can adapt to effective pollinators regardless of our expectations, and nature continually surprises us: as I was writing this chapter I spotted a study that strongly hints that at least one species of South American frog may be a pollinator.

Not all island bird pollinators are generalists undergoing interaction release, of course. Nectar specialists such as sunbirds and hummingbirds also occur on islands. However, they are part of the broader set of flower-visiting birds that we're considering in this book. To appreciate their diversity and importance, let me take you back across the Atlantic to the Caribbean.

Cuban (r)evolution

The islands that make up Cuba have one of the most impressive conservation records in the Caribbean, with about 20% of the land surface under some sort of protection. More than 370 bird species have been recorded from Cuba, including a high proportion of migrants that travel south to escape cold winters further north. There are also endemic bird species and genera that are found nowhere else, including the Bee Hummingbird, the smallest living avian species. This diversity has frequently drawn ornithologists to Cuba, including the late Jim Wiley, whose 2018 obituary describes him as 'the most important ornithologist of the Caribbean in modern times'. Before his death in 2018, Jim collaborated with my

Danish colleague Bo Dalsgaard to summarise and analyse observations he'd made of nectar-feeding birds in Cuba, and to set them within the context of theoretical island ecology.

To reiterate what I said earlier about interaction release, one of the expectations of birds on islands is that they tend to widen their feeding niche and consume a more diverse set of foods compared with the same, or closely related, birds on nearby continents. The same is true when we consider flowers and their pollinators: island plants are often more generalist than continental relatives, employing a wider diversity of insects and vertebrates to move their pollen around, or specialising in 'unusual' pollen vectors. In other words, they have wider pollination niches. In both cases the (theoretical) reason is the same: islands offer fewer resources because of their smaller size and lower abundance and diversity of plants and animals. In ecological terms, 'resources' refers to food for birds or to pollinators in the case of plants.

To investigate this phenomenon in Cuba, Bo and Jim compiled a remarkable list of bird nectar-feeding records. Intriguingly, for some of the migrant North American warblers, these are the only records of feeding from flowers in otherwise insectivorous bird species. This fits nicely with our expectations, and it's tempting to compare this to tourists trying unfamiliar foods when they holiday in exotic climes. That's a bit of a conceptual stretch, I know, but I hope you get the idea.

Not only did Bo and Jim record a surprising diversity of avian nectar feeders, they also related it to an intriguing pattern that's been discussed for some time: compared to its size, Cuba has fewer species of hummingbirds than we might expect. Only two species breed there, the Cuban Emerald and the Bee Hummingbird, whereas other similar-sized mountainous Caribbean islands typically have between three and five species. It was David Lack, again, who first drew attention to this in what was to be one of his last scientific papers, published in 1973, the year that he died. In that paper, he admitted that he

> cannot suggest why Cuba has only two species,
> and it is the more remarkable since it is one of

> the highest islands, much the largest, and the nearest to Central America (Yucatan), also it has more species of resident land birds than any other island except Hispaniola.

Bo, Jim and their colleagues thought that they had one possible answer to this conundrum: the high diversity of non-specialist nectar feeders, both migrant and endemic, might have put an ecological brake on the evolution or colonisation of further hummingbird species. It's certainly an intriguing idea, but one that's difficult to properly test: understanding why things do not occur is much more challenging than understanding why they do.

Sky islands

David Lack's point that Cuba is 'one of the highest islands' in the Caribbean leads us neatly into the final section of this chapter. As we've seen, oceanic islands are biologically important places for several reasons. They are often remote and isolated from the nearest continental mainland, to which they have never been connected, and they can occur as archipelagos in which endemic species may be restricted just to one island. They can also be highly dynamic environments, either volcanically or in terms of climatic extremes such as hurricanes and droughts. But if we define them in these biological, geological and climatic terms, then 'islands' may not be just oceanic – the concept has been broadened to include isolated mountains, as what are often described as 'sky islands'.

Mountainous areas provide us with intriguing instances of bird-flower ecology and evolution that are every bit as fascinating as those we find on remote ocean islands. For a start, like islands, some mountains support their own endemic birds. The Eastern Arc Mountains, for example, run from Kenya southwest into Tanzania. They are at least 100 million years old, the most ancient mountains in that region of Africa, and different parts of the range host highly endemic species of animals, plants and fungi. As I mentioned in Chapter 7, I was lucky enough to visit this area in 2011 with the Tropical Biology Association, and saw a lot of locally restricted species, including the Amani Sunbird that's named after the area in which we were based. Despite its name, the species has a fairly wide distribution compared to some other birds, such as Loveridge's Sunbird – which is confined to elevations of 1,200–2,600 metres in the Uluguru Mountains in eastern Tanzania. The population of Amani Sunbirds is estimated to be about 37,000 individuals, so the species is far from rare, but with such a restricted distribution it's considered endangered. Similarly, the Mountain Sunbird, as the name suggests, is found only between 1,125 and 1,550 metres in the mountains of Luzon, in the Philippines. And in New Guinea, the Long-bearded Melidectes (a type of honeyeater) occurs only on a restricted range of mountains between 2,750 and 4,200 metres. Very little has been documented about even fundamental aspects of its ecology, such as diet and breeding behaviour.

There are hummingbirds that are also constrained to high mountain areas, including the Sword-bill that I discussed in the previous chapter, which is encountered only at 1,700–3,500 metres in the Andes. Much more restricted in distribution is the Santa Marta Sabrewing, a large, spectacular blue and green species, found only in the Santa Marta Mountains of northeast Colombia. Intriguingly, it's an altitudinal migrant, and in the February–May dry season it can be located in humid forest at 1,200–1,800 metres. During the wet season from June to October, it moves up into high-elevation open páramo vegetation, up to about 4,800 metres.

Mountainous regions impose some complex constraints on the relationship between plants and their pollinators, owing to extreme weather and lower oxygen levels. Work by Robert Cruden in the early 1970s suggested that at higher elevations birds replace bees as pollinators because they are less susceptible to inclement mountain weather. Since then, several studies have supported this idea, including Agnes Dellinger's work on *Axinaea* in the Andes that I referred to in Chapter 6. However, it almost certainly depends on the types of birds and bees under consideration. To save energy, small-bodied hummingbirds at high elevation often go into a state of torpor when the temperature drops, whereas bumblebees can tolerate quite cold conditions because they are able to generate heat internally and are well insulated. When I worked

PLATE 1 A Buff-tailed Coronet feeding from the dangling flowers of a *Fuchsia* sp. (Onagraceae) in Colombia. (© Alexander Schlatmann)

PLATE 2 Malachite Sunbird feeding on the flowers of *Aloe africana* in South Africa. (© Ethan Newman)

PLATE 3 Sometimes it's easier to walk: a Blue-mantled Thornbill feeding on flowers of *Gentianella hirculus* in the high Andes of Ecuador. (© Jesper Sonne)

PLATE 4 A Rainbow Lorikeet on the inflorescence of an urban banksia tree, Coogee Bay, Australia. The strange colouration is due to the poor air quality from the bushfires around Sydney. (© Jeff Ollerton)

PLATE 5 A Cuban Green Woodpecker taking nectar from the flowers of a Geiger Tree. (© Bo Dalsgaard)

PLATE 6 Green Woodhoopoe, normally described as an insectivorous species, visiting the flowers of a coral tree (*Erythrina caffra*) in South Africa. (© Ethan Newman)

PLATE 7 A Moorland Chat pollinating flowers of a giant lobelia on Mount Kenya. Note the pollen on the bird's forehead. (© Alexander Schlatmann)

PLATE 8 A Gray-hooded Sierra Finch pollinates the endemic Argentinian cactus *Echinopsis leucantha*. (© Pablo Gorostiague)

PLATE 9 The bird-pollinated southern African shrub *Melianthus comosus* (also known as *M. minor*) produces black nectar. (© Joey Santore; youtube.com/c/CrimePaysButBotanyDoesnt)

PLATE 10 A Glittering-bellied Emerald hummingbird pollinating the flowers of *Cleistocactus baumannii* (Cactaceae) in Argentina. (© Pablo Gorostiague)

PLATE 11 So-called 'Indian paintbrushes' (*Castilleja* sp.) stand out like small campfires on the dry, rocky mountain slopes of Colorado, USA. (© Jeff Ollerton)

PLATE 12 A Ruby-crowned Tanager visiting the flowers of *Scybalium fungiforme* in the Atlantic forest of Brazil. (© Felipe Amorim)

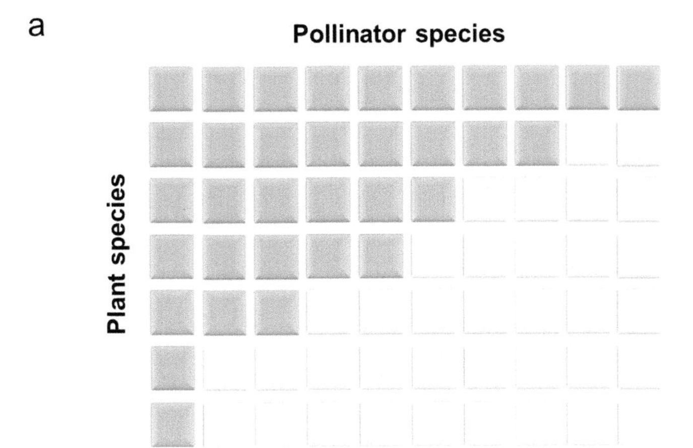

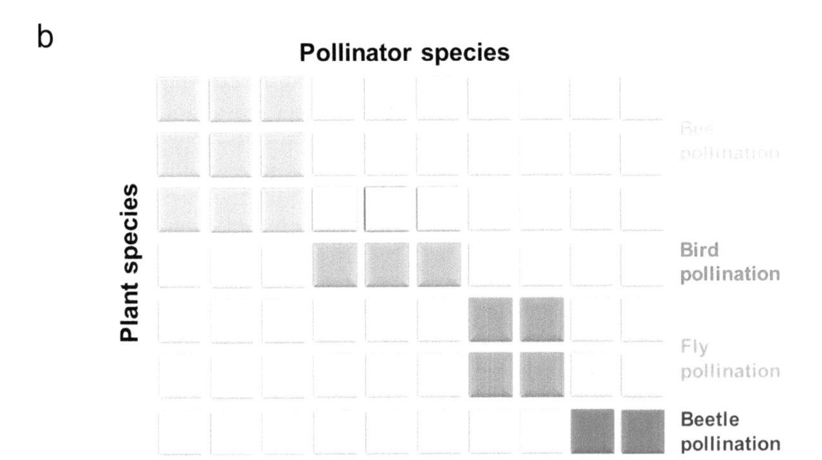

PLATE 13 The non-random nature of plant–pollinator interactions: (a) nestedness; (b) modularity. Refer to the text of Chapter 9 for further details.

PLATE 14 a and b The bird-pollinated plants of Tenerife are a distinctive and spectacular part of the flora of the Canary Islands. At the top is Canary Island Foxglove, which is solely bird-pollinated. Next, the author admires the towering inflorescences of the Mount Teide Bugloss, pollinated by both birds and insects. (© Jeff Ollerton)

PLATE 15 A Common Chiffchaff is one of the bird pollinators of *Anagyris foetida* (Fabaceae) in southern Spain. (© Francisco J. Valtueña)

PLATE 16 Pollen on the face of a Eurasian Blue Tit in the early spring. Bird mist-netted under licence in Northamptonshire, UK. (© Lynne Barnett)

PLATE 17 A Nazca ceramic with different species of hummingbirds visiting a flower. (© Jeff Ollerton)

PLATE 18 Inkwash painting of 'Flowers and Bird' by Miao Jiahui (1841–1918), Qing dynasty. (Courtesy of Zong-Xin Ren)

PLATE 19 'Pomegranate flowers and Warblers' by Lu Ji (1439–1505), Ming dynasty. (Courtesy of Zong-Xin Ren)

PLATE 20 The pickguard of a Gibson Hummingbird guitar. (© Jeff Ollerton)

PLATE 21 My friend Henrik playing his Paul Reed Smith guitar. Note the inlayed hummingbird fret marker. (© Jeff Ollerton)

PLATE 22 Wine label showing a Rufous Hummingbird visiting the flowers of a hibiscus. (Courtesy of Nick Waser)

PLATE 23 The Humboldt Brewing Company in California uses a hummingbird as its logo. This bottle label for their Red Nectar ale features a Rufous Hummingbird visiting flowers. (Courtesy of Nick Waser)

PLATE 24 A South African wine label depicting a Cape Sugarbird and a pincushion (*Leucospermum* sp., Proteaceae) inflorescence. (© Jeff Ollerton)

PLATE 25 Scarlet Honeyeater visiting flowers of a species of *Grevillea* (Proteaceae) in Australia. (© Geoff Park; www.naturaldecisions.com.au)

PLATE 26 A Scintillant Hummingbird pollinating the flowers of an orchid (*Arpophyllum giganteum*) in Costa Rica. (© Adam Karremans)

PLATE 27 A Mariqua Sunbird forages for nectar from the flowers of a species of *Grewia* (Malvaceae) in Kenya. (© Alexander Schlatmann)

PLATE 28 A Noisy Miner defending the nectar-rich flowers of *Banksia robur* from other birds, in Queensland, Australia. (© Kit Prendergast)

with Stella Watts in the high Andes of Peru we were seeing bees at almost 4,000 metres, and bumblebee expert Paul Williams has documented them at even higher elevations in the Himalayas. On the other hand, when I visited Nepal in March 2019, during a cold, snowy trek up to over 3,650 metres in Langtang National Park, almost the only flower visitors that we saw were birds. A diversity of species, including Fire-tailed Sunbird, Coal Tit, several leaf warblers and some unidentified rosefinches, were seen visiting, and almost certainly pollinating, the flowers of large rhododendron trees. The spectacular Fire-tailed Sunbird, incidentally, is another mountain migrant that moves from below 1,000 metres in the winter to above 4,000 metres in the summer.

In total we recorded 13 different birds taking nectar from the flowers of three large rhododendrons in Nepal, some of which are substantial trees reaching 20 metres in height. The flowers of these species vary from pure white to deep red, but birds do not discriminate between them. When I wrote up our observations with my Nepalese colleagues we pointed out how important these rhododendron forests are for high-elevation communities in that country. The local Sherpa and Tamang peoples rely on the forests for a range of goods and services, including wood for fires, timber for construction, animal forage, beekeeping, as well as medicinal and culturally important plants. It's also a place of pilgrimage, religiously for Hindus and Buddhists, and secularly for birders and other international visitors who add to the growing income from tourism. Yet other than a few scattered observations such as ours, virtually nothing is known about the role of birds in maintaining seed production, and therefore long-term persistence, of the rhododendron populations.

Nepal is not alone. Other mountainous regions are similarly neglected. The few studies made of birds as pollinators in New Guinea, for example, have focused on the lowlands rather than the much more diverse, though less tractable, uplands. The pollinating roles of species belonging to families endemic to those islands and mountains, such as the berrypeckers and longbills, have hardly been studied by scientists. Whether the indigenous peoples of New Guinea have their own knowledge about these birds and their flowers is also unknown, as far as I am aware. This is a topic that I will return to in Chapter 14.

As fascinating, exciting and enigmatic as islands are, it's time to leave them and move back to a continent. Considerations of the Canary Islands 'Tertiary relict' flora, the work of European researchers such as Charles Darwin, David Lack and their intellectual inheritors, brings us to an important question when it comes to bird–flower interactions: why is Europe so weird?

CHAPTER 13

The curious case of Europe

Writing in the early 1960s, Dutch scientist Bastiaan Meeuse (1916–1999) admitted that the first time he encountered a hummingbird in North America, he mistook it for a hawkmoth, 'as any European greenhorn would have done'. If this came as a surprise to his American readers, he added, they must understand that

> In all Europe, there are no birds which are interested in nectar or pollen for food … As a result, there has always been a tremendous emphasis on insects in European pollination work, and the birds have been downright neglected.

That statement, from Meeuse's 1961 book *The Story of Pollination*, summed up a notion that had long persisted and which was widely regarded as common knowledge – that there were no native bird-pollinated plants in Europe. Meeuse himself had spent the first 15 years of his life in Indonesia, where, according to an obituary, 'the beauty and variety of the plants he saw triggered Meeuse's awareness of the natural world, and set him on course to become a biologist'. It is therefore inconceivable that he was not familiar with flower-visiting birds, and his statement about being a 'European greenhorn' doesn't quite ring true. His assessment that 'birds have been downright neglected' as pollinators was, however, not an exaggeration, at least as far as the European literature was concerned.

The first great collation of the available science relating to flower pollination was German scientist Hermann Müller's 1873 volume *Die Befruchtung der Blumen durch Insekten* ('The pollination of flowers by insects'). The title comes as no surprise: insects were what Müller was used to observing. Müller's book was subsequently brought up to date and translated into English, with a preface by Charles Darwin, as *The Fertilisation of Flowers* (1883). The text of this edition contains occasional references to birds as pollinators – fuchsia flowers visited by New Zealand Bellbirds, for example, and hummingbird pollination of several other flowers. However, there is also a footnote on page 24 that 'birds may be left out of the question, as they do not play an important part in fertilising any native plants', again reinforcing the idea that Europe is the odd-continent-out. Given the state of knowledge at the time, this was a reasonable comment. Less understandable is why the main index of pollinators is solely an 'index of insects'. Birds get no entry – though they do appear, collectively as 'birds', in the 'index to bibliography', with just nine sources listed under that heading. This is perhaps all the more surprising as Hermann Müller's brother Fritz lived for much of his life in southern Brazil, he is mentioned numerous times in the book, and the two corresponded frequently about pollination and wrote about hummingbird flowers. In August 1871, Fritz sent a letter to Hermann regarding the species in his garden:

> A large and splendid hummingbird, whose black breast glows like a red-glowing coal when he is aroused in any way, has, with his inconspicuous little wife, claimed for himself almost complete rule over the Abutilon species in my garden, and chases away all other species. All uncovered blossoms are fertilised by him.[1]

As we saw earlier, territoriality of good nectar sources amongst male flower-feeding birds is common, and this is quite an early account of such bird behaviour. But why did Hermann not include

---

1 I'm grateful to Nick Waser for providing a more accurate translation of this from the original German.

it in his book? I do wonder whether he simply did not consider birds of particular importance, beyond tropical amusements, perhaps following Darwin's lead. Müller had corresponded with Darwin, who had encouraged his research on pollination. Darwin himself, writing in his 1876 book *The Effects of Cross and Self Fertilisation in the Vegetable Kingdom*, had relegated bird pollination to a footnote on page 371 in which he stated that he would 'give all the cases known to me of birds fertilising flowers'.[2]

Matters were much improved by the appearance of a second great compendium of literature on pollinators and pollination. Paul Knuth's *Handbook of Flower Pollination* was based on Müller's book but expanded it greatly, resulting in three volumes, initially in German. The English versions, translated by J. R. (James Richard) Ainsworth-Davis, were published between 1906 and 1909, and widely reviewed in journals, including *Science*. Knuth summarised what was then known about bird pollination in a section just over four pages in length entitled 'Plants with bird-pollinated flowers, Ornithophilae'. This section begins with the following text:

> Plants in which the flowers are pollinated by birds (humming-birds, honeysuckers, rarely woodpeckers) are found especially in the tropics.[3]

Almost a century later, writers such as Meeuse were making similar statements, that bird pollination was rare outside of tropical habitats, and in particular that there was an absence of bird pollination in Europe. This echoed down the decades into the new

2 I can't resist writing a footnote about a footnote ... One of those cases was that Darwin had 'been assured that at the Cape of Good Hope, Strelitzia is fertilised by the Nectarinidae [*sic*]'. That assurance was wrong, or at least certainly not the whole story. As Gareth Coombs and Craig Peter have shown, and the cover of this book illustrates, *Strelitzia reginae* is pollinated by generalist passerines such as the Cape Weaver. Sunbirds are nectar thieves.

3 Ainsworth-Davis's translation renders this as sparrows (*Spatzen* in German) but the original states *Spechten* – woodpeckers. Again, I'm grateful to Nick Waser for correcting this.

millennium. The problem, as is so often the case with common knowledge, is that it was wrong.

How science works (and how it doesn't)

In late 2004 I was sent a manuscript to review by one of the editors of the journal *Oikos*. Receiving such requests is an approximately weekly occurrence for most publishing scientists once they establish themselves within a field. The 'peer review' of scientific papers is an important dimension to the modern scientific process: once a piece of research has been written up as a manuscript, it is submitted to a journal and then assigned to one of the journal's editors. That editor in turn decides who to send it to for review: usually two, but sometimes more, scientists studying similar topics. Most of us receive far more requests than we can ever agree to, but a rule of thumb is that, for every research paper that one publishes, a good citizen of science should review at least two manuscripts. In practice the distribution of reviewing effort is actually bimodal: a small proportion of scientists review a disproportionately large fraction of manuscripts, whilst a second group in effect parasitise the system and review next to nothing. I try to be a good citizen and do my fair share, though this very much depends on what else is happening in my life at the time. In the case of the *Oikos* manuscript, I did what I always do – I scanned the title and then read the abstract, before making a decision. That decision did not take long, and I immediately agreed to review it. How could I not? The title alone was enough to entrap me:

> 'First confirmation of a native bird-pollinated plant in Europe'

The manuscript had been written by Spanish scientist Ana Ortega-Olivencia and a team from the University of Extremadura. Ana and her collaborators had studied a small tree in the pea and bean family called *Anagyris foetida*. The tree is native to the Mediterranean and it's thought to be one of those 'Tertiary relict' species that I mentioned in the previous chapter, a species that has persisted in the region for the last several million years or so and reflects a type of vegetation that has now largely disappeared. There are quite a number of these plant species across southern Europe, hanging on in low numbers in often inaccessible places, despite significant changes in climate since they first flourished in Europe. Ana's group spent two years observing and carrying out experiments in a couple of populations of *Anagyris* in southwestern Spain. This field work included counting flower visitors, analysing nectar, and measuring flower traits, looking at how much pollen each type of visitor was carrying, and finally which visitors were the most effective at depositing that pollen. Their conclusion: *Anagyris foetida* is bird-pollinated, the first time that this had been demonstrated for any native European plant (Plate 15).

The three most important pollinators of *Anagyris* in terms of how frequently they visited flowers and their ability to transfer pollen were Common Chiffchaffs, Eurasian Blackcaps and Sardinian Warblers. There were few insect flower visitors, despite the presence of significant amounts of nectar in each flower, which the researchers explained as being due to the tree flowering during the Iberian autumn and winter, when the weather is cold, rainy, windy and misty. That's weather that is all but guaranteed to deter insects but which can be tolerated by these birds. The flowers are unusual for a legume, but seem to fit well with their reliance on birds as pollinators: they are scentless, pendulous, tubular, and have no lip on which insects can land. They are also, to our eyes, a rather dull greenish-yellow with a brown patch on the upper petal, not at all what we might expect a bird-pollinated flower to look like. When I spoke with Ana recently she told me that she got the idea of observing this plant more closely because its combination of odd flower form, large amounts of nectar and late flowering meant that it was 'very different from the rest of the European legumes' that she was studying at the time.

The Swedish naturalist Carl Linnaeus, founder of the binomial system used in taxonomy today, described *Anagyris foetida* in the eighteenth century. The birds that pollinate it likewise have long scientific histories and are familiar species. So why had it taken so long to discover this important interaction? The answer is simply that no one had looked. Perhaps because 'everyone knows' that there are no bird-pollinated plants in Europe, it wasn't worth searching for them, a classic case of blinkered science if ever there was one.

Ana's research was exciting work and such a pleasure to review, one of a handful of times that I've read a manuscript and thought: 'wow, this really challenges our understanding of the topic'. I described another example in Chapter 6, when I discussed Anton Pauw's work on bird-pollinated milkweeds, but this is not an everyday (or even every-year) occurrence for a scientist. The most recent example, for me, was reviewing a manuscript about a seaweed 'pollinated' by a crustacean. I've used quotation marks because seaweeds are algae and they don't have pollen; nonetheless, discoveries like this dramatically challenge our understanding of the biodiversity of species interactions, how ancient these sorts of relationships actually are, how they evolve, and their ecological importance.

Meanwhile, back on land, and not satisfied with just working on *Anagyris*, Ana assembled a team which went on to discover bird pollination in another set of European plants, the figworts, including at least one Canary Island species. Nectar in some of these species is hexose-dominated, as expected for flowers with generalist bird pollination systems.

Historical precedents

The really surprising thing about the findings of Ana and her team is that *Anagyris* and the figworts are European species that are clearly specialised for pollination by nectar-feeding birds, sometimes in concert with insects. What is not surprising is that these European birds visit the flowers. There have been plenty of previously published observations, over the last 100 years, showing that small passerines in Europe probe flowers for nectar. It's also not uncommon for migrant warblers to arrive in their northern breeding range loaded with substantial amounts of southern European pollen on their faces (sometimes called 'pollen horns') and bodies. Likewise, resident tits are sometimes mist-netted by bird ringers and found to be dusted with the pollen of early spring-flowering willows (Plate 16). Evidence suggests that this is an important part of their diet early in the year. In the 1980s, British biologist Quentin Kay studied the behaviour of Eurasian Blue Tits feeding on the nectar of willows in Wales. Based on some very detailed observations, he calculated that the tits could obtain their total daily energetic requirements by foraging on willow catkins for between three and four hours. Furthermore, he was convinced that the birds played a significant role in the pollination of these flowers. He may well be correct, but at the time of writing it's not been fully tested. It's a set of experiments just begging to be properly carried out.

The relationships between tits and other trees in Britain is even less well understood. In early spring sunshine I have often watched in fascination as Eurasian Blue Tits work their way meticulously along the branches of Cherry Plum, a favoured, early-flowering tree that's a longstanding introduction to Britain. They quickly dip into flowers as they forage, stimulating my imagination. Are these birds looking for nectar? Or just insects? Or both? Do they damage the flowers, or could they perhaps be pollinating them? Such familiar birds and such a common-place tree – why do we know so little about how they interact? This topic also interested Charles Darwin and, writing in the journal *Nature* in 1874, he discussed how birds destroyed the flowers of Common Primrose but did less damage to those of cherries, and that nectar seemed to be the primary motivation for their probing visits.

To bring together the scattered publications on the topic of nectar feeding in European birds, in 2014 a review was published by a joint Portuguese/Danish/Spanish team led by Luís da Silva. Their scrutiny of the literature uncovered a total of 62 sources, including mentions by Erasmus Darwin, grandfather of Charles, and other eighteenth-century naturalists. Gilbert White, in his book *The Natural History and Antiquities of Selborne*, published in 1789, writes:

> A rare, and I think a new, little bird frequents
> my garden, which I have great reason to think
> is the pettichaps: it is common in some parts
> of the kingdom; and I have received formerly

> several dead specimens from Gibraltar. This bird much resembles the white-throat, but has a more white or rather silvery breast and belly; is restless and active, like the willow-wrens, and hops from bough to bough, examining every part for food; it also runs up the stems of the crown-imperials, and, putting its head into the bells of those flowers, sips the liquor which stands in the nectarium of each petal.

'Pettichaps' is an old English word applied to several species of warbler, but here White seems to be referring to the Garden Warbler drinking nectar from the Imperial (or Crown) Fritillary, an Asian plant that is almost certainly bird-pollinated in its native range (see Chapter 14 for a discussion of the strong cultural associations of this plant). Despite Bastiaan Meeuse's dismissive comment, European gardeners frequently see warblers and tits visiting the flowers, and experiments have shown that the birds will pollinate them in a garden setting. So this is a concrete and historically well-known example of birds fitting into an existing pollination network, in ornamental gardens, that the scientific community assumed did not have bird–flower interactions. It may even hint at underlying links with native flowers that teach the birds to visit flowers, though it could be that a tendency to explore flowers is just a deeply ingrained aspect of bird behaviour and cognition.

As for Erasmus Darwin, a search for what he had to say led me into a bit of a digital maze,. The authors of the review had referred to his 1791 'poem in two parts' *The Botanic Garden*. To track down exactly what Darwin senior had written about birds visiting flowers in Europe, I turned to the ChatGPT artificial intelligence (AI) system, which at the time was a bit of a novelty and generating a lot of media interest. I asked the AI: 'What did Erasmus Darwin say about birds visiting flowers?' This is the answer that came back:

> Erasmus Darwin was a naturalist, physician, and poet who wrote about a wide range of topics, including birds and flowers. In his 1794

> poem 'The Loves of the Plants,' he describes how different species of birds visit flowers in search of nectar:
>
> *Birds of each hue, from the gay parrot's beak*
> *To the goldfinch, painted in tiny streak,*
> *Sip nectar from the cups that bloom and bow*
> *With every breeze, and nod and dance below.*

Now, the first part of this, about Erasmus Darwin's life, is true. It's also true that he wrote a poem with that title, though it was published in 1791, not 1794, as the second part of the 'poem in two parts'. My initial reaction was that this was a quite fascinating and rather early literary reference to birds feeding from flowers. But when I double-checked those lines against the published version of the poem on the Project Gutenberg website, I discovered that they did not exist. ChatGPT had made them up! As a scientist and writer I try to be rigorous when quoting and citing sources, and if I hadn't double-checked this I might have taken it at face value. And one could argue that my initial question – What did Erasmus Darwin say…? – could have been more precisely phrased. Nonetheless, I was not expecting an AI to try to pass off as fact something that was so demonstrably untrue.[4]

What did Erasmus Darwin really say about birds visiting flowers? Actually not that much with regard to Europe, simply:

> A bird of our own country called a willow-wren (Motacilla) runs up the stem of the crown-imperial (Frittillaria coronalis) and sips the pendulous drops within its petals.

It's a little unclear, but he seems just to have misread this in White's book rather than making any observations himself. Erasmus actually has more to say about hummingbirds, suggesting for example

---

4 As Canadian ecologist Stephen Heard notes on his blog, ChatGPT has many uses in scientific writing, but providing accurate information is not one of them.

(incorrectly) that the inflated flowers of *Cypripedium* orchids mimic bird-eating spiders, which 'seems to be a vegetable contrivance to prevent the humming-bird from plundering its honey'.

Back to that review by Luís da Silva and colleagues. From these early observations onwards, the team revealed a surprising diversity of birds (46 species) and of the flowers that they visit (95 species, less than one-third of which were not native to Europe). That study alone added three families to the table of bird diversity in Chapter 3 – long-tailed tits, Old World orioles, and accentors – for which we otherwise have no records of flower visitation. So it's clear that we have known for a very long time that European birds will visit flowers. It seems, however, that there are few European plants that specialise in these birds as pollinators. And that raises an interesting question.

Why is Europe so odd?

Given that in Europe there is a handful of specialised plants that are pollinated by generalist passerine birds, and that these birds may play a role, along with insects, in pollinating some early-flowering trees and shrubs, the bigger, biogeographic question is: why is bird pollination so rare on this continent? I have to confess, it's a real puzzle, and there's no easy answer. As with so many conundrums in ecology and evolution, there may be multiple factors that have come together to result in the patterns that we see.

One thing is certain, though: the lack of specialised bird-pollinated flowers is not due to a lack of birds, at least not those generalist birds that, as we have seen, can be effective pollinators and drive floral evolution. Tits and warblers that visit flowers are certainly common enough, even if some species have declined in abundance as Europe has become increasingly industrialised and intensively farmed.

One clue to this conundrum may be found in the description of *Anagyris* and some other plants as 'Tertiary relicts'. It's certainly possible that as Europe's climate has changed over the course of the last few million years, so we've lost many of the plants which were bird specialists. In this scenario, the few bird-pollinated plants that hang on in Europe are now the remnants of a flora in which bird pollination was once more prevalent. That can't be the whole answer, though, because natural (as opposed to anthropogenic) climate change has been a feature of all parts of the world over this time span, and bird pollination is far more common in temperate zones elsewhere.

Another, potentially more significant factor, I think, is that there are no specialist flower-feeding birds in Europe, or indeed in Asia north of about 40 degrees of latitude in the case of the sunbirds. Some of the white-eyes can be found at higher latitudes in eastern Asia, including parts of Japan and China, but otherwise Eurasia is characterised by a paucity of specialist flower feeders. This fact is in stark contrast to North America, where hummingbirds, and specialised hummingbird flowers, are found even in Canada and southern Alaska. Likewise, in South America one species of hummingbird, the Green-backed Firecrown, can be found as far south as Tierra del Fuego. In both hemispheres, specialist birds and their flowers are found above 50 degrees of latitude, and of course the birds can migrate to warmer climes during their respective winters, whilst the plants go into dormancy as adults or as seeds.

In comparison, sunbirds only ever undertake relatively short elevational and latitudinal migrations, and it is likely that they cannot tolerate the colder winters further north. The closest that any of the specialist birds get to Europe is in the eastern Mediterranean, where the Palestine Sunbird is commonly encountered as far north as Lebanon and southern Syria. Why, though, are these birds not found in southern Europe, where other cold-intolerant migrants pass the winter months? Again, perhaps sunbirds lived in this part of Europe several million years ago, then subsequently became extinct. But if that's the case their fossils have not been found. Indeed, the hummingbird relative I discussed at the start of the book is the only (presumed) flower specialist to have been so far discovered in Europe. However, until we find further fossils of this enigmatic and unexpected species, with pollen in or on its body, we will not know for certain.

The scientists and naturalists who for decades wrote about the absence of bird pollination in Europe were only partly correct, because the birds and the flowers were there, they just weren't what they were expecting. Having said that, the current absence of specialist flower-feeding birds, and especially the paucity of bird-adapted flowers, means that Europe, and indeed northern Eurasia, remains a biogeographical mystery. Where are those specialised birds? Why are they not here? Were they ever here? And what about the plants? Why have those bird-specialist flowers not evolved? Or have they, and are they hiding in plain sight, waiting for observant naturalists to say 'hang on a minute, something unusual is happening here'? Perhaps some of those willows I referred to earlier will prove to be more bird-adapted than we've realised.

Europe is indeed a curious case, as far as birds and flowers are concerned, when we compare it to the other continents. Gaps in our understanding of the natural world are always frustrating, because many of us have an innate desire to understand how nature functions and why things are as they are and not some other way. But on the bright side, there's no doubt plenty more exciting stuff to discover. Perhaps now is the time to reflect on the origins of our knowledge, to pause and consider: how did we come to know so much about birds and their interactions with flowers, and where did it all begin?

CHAPTER 14

'After the Manner of Bees'

Just north of the Danish capital city of Copenhagen, nestled against the lapping waters of the Øresund, lies the Louisiana Museum of Modern Art. Louisiana is one of the most interesting art centres in Scandinavia, a complex of interconnected buildings, open spaces and underground galleries that fascinate visitors for hours. I first went there in 2016 with Danish scientist Bo Dalsgaard, whom you met in earlier chapters, and my wife Karin. As we meandered through the site, pausing to view sculptures and installations, I spotted an exhibition of early South American ceramics. This is part of a collection donated to Louisiana in 2001 by Danish actor, dancer and choreographer Niels Wessel Bagge (1908–1990), and it consists of 450 pre-Columbian artefacts from across the Americas. One of the items in particular caught my eye: a tall, oddly shaped, 'stirrup spout' ceramic pot decorated with hummingbirds.

The artist who painted the pot had arranged seven birds in a wheel formation, with their beaks all entering the same flower (Plate 17). I'd seen hummingbirds on indigenous New World ceramics before – they are a common motif in Peruvian Nazca pottery – so I took a couple of photographs with my phone and moved on.

The Nazca culture that created these ceramics flourished on the dry southern coast of what is now Peru, between about 100 BCE and 800 CE. Hummingbirds clearly had enormous significance to these people – quite literally, judging by the 93-metre-long image they etched into the desert. The 'Nazca Lines' are amongst the most famous of all South American archaeological sites, and several of the geoglyphs are iconic and easily recognisable, including a monkey with a spiral tail and a very stylised but clearly recognisable hummingbird. That hummingbird has an extremely elongated bill that seems far out of proportion, and a figment of the artists' imaginations. Perhaps it's meant to represent the Sword-billed Hummingbird? The proportions are certainly not far off, with a bill longer than the body of the bird being its unique trait. However, Sword-bills do not occur in that part of Peru – they are an Andean species, found only above 1,700 metres elevation. Does this indicate that there was contact between the Nazca people and those of the high mountains? Or perhaps that bird was found at lower elevations in the past, when the climate was wetter? Or maybe this is just a general representation, not meant to depict any one species?

In 2019 a Japanese team led by Masaki Eda published a research paper entitled 'Identifying the bird figures of the Nasca pampas: an ornithological perspective'. They concluded that the well-known hummingbird geoglyph, and possibly two others, depicted hermits rather than any of the other types of hummingbird, on the basis of the tail morphology, noting that 'In Peru, long and pointed tails only occur in hermits ... whereas the tails of typical hummingbirds are forked or fan-shaped.'[1] This is an interesting idea but it relies on the tail being an accurate representation of a species or group of species. And if it is accurate, what about the bill? No living hermit hummingbird species has a bill longer than its body, excluding the tail. In fact with the exception of the Sword-bill, no other bird, of any kind, has such a beak. Could the Nazca hummingbird actually represent an extinct species? Unless preserved specimens are found, we'll never know – though that is not as unlikely as it sounds: the dry, stable climate of this region has mummified humans, so perhaps preserved birds are also out there somewhere?

Elsewhere in the Americas, hummingbirds also had, and continue to have, huge cultural significance for indigenous peoples. The Aztec god Huītzilōpōchtli is depicted as a hummingbird or as a man dressed as a hummingbird, and as a deity he represented the sun, war and human sacrifice. Warriors who died in battle

1 See Chapter 3 for an overview of hummingbird classification.

transformed into hummingbirds and joined Huītzilōpōchtli. The association of hummingbirds with the sun seems reasonable – they are vivid reflectors of sunlight – but war and sacrifice are less obvious to me. The Aztecs used feathers from a wide diversity of birds to produce stunning headdresses, fans and items of clothing, often with ceremonial functions or as symbols of status. Hummingbirds feature regularly in the pieces that survived the Spanish conquest of Mexico, and still retain their original iridescent colours, testament to it being the feather structure rather than pigment which gives them their hues. After the Spanish imposed their religion on the peoples of this region, the professional featherworkers shifted from producing traditional items to objects of Christian significance. Featherwork panels showing religious scenes, and even a bishop's mitre, all include hummingbird feathers. The original spiritual significance of the hummingbirds was of course lost, though their beauty remained.

In the present day, hummingbirds are still revered by indigenous Americans. They play a role in shamanic ayahuasca rituals amongst the Shipibo-Conibo people of Amazonia and in the Pueblo culture of southwest USA and Mexico, for example. Trying to understand the motivations and knowledge of these people's ancestors is not a simple task, however.

Sometimes familiar things surprise us the most

One of the photographs that I took that day at Louisiana was used in the opening chapter of *Pollinators & Pollination*, where I described it simply as: 'Hummingbirds visiting a flower on a piece of pre-Columbian South American pottery'. But had I really looked at that image and considered what the pot was telling me? Had I thought deeply about what I was seeing, and what the painter of this ceramic object was trying to depict? It was only when I revisited the image while contemplating this chapter, and the relationships between indigenous peoples and flower-visiting birds, that I truly saw the piece for what it was: a story of ecology and diversity coming to me from an unfamiliar, pre-colonial culture of the distant past. Why had I not truly noticed it before? This image was different to most other Nazca hummingbird ceramics that I've seen: it depicted not just the hummingbirds, but the ecological interactions between birds and flowers. That in itself was unusual, but then I noticed something truly surprising about the birds. Some had short bills, others were very long. The bills varied from dead straight to decidedly curved. The colours and patterns of the plumage varied too, as did the lengths of the tail feathers. Look again at Plate 17: each bird is a different species.

The word 'ecology' was only coined in 1866, 'taxonomy' a little earlier, in 1813. But here was a Nazca artist painting a pot more than 1,000 years ago who was expressing themselves both as an ecologist, showing the interactions between flowers and hummingbirds, and as a taxonomist, documenting the differences between those birds. What of the flower? Was the artist stating, in polychrome pigments, 'these birds visit flowers'? Or was it a different ecological statement: 'these birds all visit this *particular* flower'? We'll never know, because there's not enough detail on the flower to say for certain what it is. It has eight petals, and there are few South American plants that routinely have their petals in sets of eight, so perhaps it just represented 'flowers'.

It's clear that American indigenous cultures knew about birds as flower visitors prior to colonial contact, even if much of that knowledge is now fragmentary or lost, and we do not know the depth of their understanding of the process of pollination. As well as hints from ancient painted ceramics, we have present-day accounts passed down, largely, from oral histories. But what about the rest of the world? Where do flower-visiting birds fit in other cultures?

Beyond the Americas

In 2014 the Intergovernmental Platform on Biodiversity and Ecosystem Services (IPBES) organised a workshop in Panama entitled *Indigenous and Local Knowledge about Pollination and Pollinators Associated with Food Production*, and a report from that event was published by UNESCO in 2015. As you might imagine, most of the indigenous and local knowledge related to bees and other insects, but pollinating birds also made their presence felt.

One of the contributions to the workshop was from James (Tahae) Doherty and Kirituia Tumarae-Teka, who related how the indigenous Māori people of Aotearoa/New Zealand recognised several bird species as pollinators or florivores and as a source of meat that they could hunt. These were the *Kererū* (New Zealand Pigeon), the *Kōkō* (Tui, an endemic honeyeater) and *Pihipihi* (Silvereye, a once uncommon species that became much more abundant in New Zealand in the mid-nineteenth century). The *Kererū* has a particular cultural importance to the Māori, and the authors described how it visits flowers of *Kōwhai* trees (species of *Sophora* in the pea and bean family). The birds pluck off the petals and eat the anthers and other reproductive parts, but not every time, and so it's possible that some flowers get pollinated this way. Although New Zealand ecologists consider it to be just a florivore, James noted that 'I have observed *Kererū* with pollen on its beak and mouth and feathers around the head'. The role of this bird as a potential pollinator certainly should be looked at more closely.

Of the other flower-visiting species, in the words of the authors, 'birds such as the *Kōkō* and *Pihipihi* were recognised by the elders as having a much clearer role in pollination. *Kōkō* were known to pollinate native trees with tubular-shaped flowers … but also introduced shrubs and trees.'

This last point is important: all cultures evolve, and indigenous cultures are not preserved in pre-colonial amber, they adapt as circumstances change. This includes integrating newly arrived species into their cultural fabric. The authors of that account in fact present an interesting example of this. Collecting honey from feral Western Honey Bee nests was considered an important seasonal event that could be tracked and predicted from the flowering of different native trees. These bees were introduced from Europe into New Zealand only in 1839 and likely took decades to become established in the wild. So Māori use of this sweet resource, although important to them, is not an ancient tradition.

But I digress. Let's get back to birds. Māori accounts describe flocks of *Kererū* see flying in the first half of the twentieth century that were so large that they would 'shade out the sun like a cloud passing over' and break the branches of trees in which they roosted.[2] From the 1960s onwards these flocks were a fraction of their former size, a result of multiple factors: competition with and predation from alien invasive species, destruction of their forest habitat, and lack of food. Even if these birds are not pollinators, they certainly are fruit eaters and seed dispersers, and therefore an important element within the ecology of New Zealand. As the authors of the workshop report note, the Māori 'recognise the collectiveness in their world – everything is connected in their culture … and provides the strength to the ecosystem. As these elements are eroded, so is the resilience and integrity of the environment and the tribe.' As we saw earlier in the book, this chimes exactly with how scientists consider the networks of interacting species and their reliance on one another to form functioning ecosystems.

Like the Māori, the ancient Egyptians were also astute observers of natural history. Tomb walls painted thousands of years ago show the flora and fauna of the Nile Valley with detail and precision: every leaf is lifelike, hairs and feathers are individually painted. Even identifiable insects are rendered, such as the depiction of the African Monarch butterfly in the scene 'Nebamun hunting in the marshes', currently housed in the British Museum. As well as being a source of food, birds wild and domesticated held a particular religious and ritualistic significance for these people, and species such as the African Sacred Ibis and Egyptian Vulture were regularly shown on reliefs, pottery and jewellery. They were also ritually mummified, along with other animals. Ornithologists with an interest in ancient Egypt have identified with some certainty more than 210 bird species from their painted, carved or sculpted representations. This is about half of the birds that are currently resident or migrant in the country. But amongst all the ducks, geese, kingfishers, raptors and small passerines that the Egyptians portrayed, there's a notable absence of sunbirds. Two species – the Palestine Sunbird and the Nile Valley Sunbird – are common in Egypt, and a third species (the Shining Sunbird) is rare or occasional. The males of all three birds are striking and colourful and, of course, they

2 This reminds me of early accounts of the Passenger Pigeon in North America – see Mark Avery's fascinating, tragic book *A Message from Martha*.

regularly visit flowers. Why did this not fire the imaginations of the ancient Egyptians in the way that hummingbirds did for cultures in the Americas? It's a mystery to me. Egyptians held bees in high reverence, and they also knew the basics of plant reproduction – the earliest depictions of pollination involve Nile Valley farmers shaking pollen from male date palms onto the inflorescences of female trees, a practice that persists to this day. Perhaps archaeologists will discover sunbird representations in the future, but at the moment we can say with some certainty that, if they exist at all, they are not common.

Elsewhere across the Middle East and Asia we encounter occasional cultural references to flower-visiting birds or bird-pollinated flowers. The Imperial Fritillary that so interested Gilbert White has long been revered in Iran, where it's a native plant, with many historical, religious, folkloric and mythological associations. The downward-pointing flowers are interpreted as the plant bowing its head in respect following the death of an important person from mythology or religious tradition, with the abundant nectar drops representing tears of mourning. Polish pollination ecologist Katarzyna Roguz and her co-workers have studied the different flower traits associated with insect and bird pollination in the fritillaries, and have come to the conclusion, based on flower colour and nectar characteristics, that there's a very good chance that bird pollination has evolved several times in the genus. For the Imperial Fritillary we're lacking observations in its native range that would confirm exactly which birds are involved, but it's likely to be small generalist passerines. Intriguingly, a Persian name for this plant translates as 'flowers of six feathers' which may just refer to the fact that each bloom is composed of six tepals that are sort of feather-shaped,[3] though I wonder whether it's also an ancient reference to its pollination system.

In Chinese art there's a long tradition of depicting birds and flowers in the same painting. It's referred to as *huāniǎohuà* (花鳥畫) and dates to at least the tenth century. These paintings,

3 Tepal is a term used when it's not possible to distinguish between a flower's petals and sepals.

though botanically and ornithologically often very accurate, were intended to be symbolic expressions of ideas and virtues as much as beautiful portrayals of the natural world. While researching this book I looked at a lot of *huāniǎohuà* to see if they ever depicted interactions between the birds and the flowers. Drawing a blank, I reached out to my colleague Zong-Xin Ren at the Kunming Institute of Botany to ask his opinion. Zong-Xin agreed with me: depictions of birds visiting flowers were extremely rare. But they do exist, and he sent me a couple of examples that he had found (Plates 18 and 19). One of these, of birds and pomegranate flowers I will return to in the next chapter.

Ancient and indigenous peoples in some parts of the world were clearly well acquainted with those birds that visited flowers. They integrated the birds, and sometimes their interactions with blossoms, in their art and culture, mythos and religious beliefs. Although they generally left no written accounts, they produced not only many images, but also oral stories that in some cases persist to this day. For example, my Nepalese colleague Narayan Koju tells me that the Lepcha people of the Himalayas named the species that we call the Fire-tailed Myzornis *Lho sagvit-pho*, which translates as 'the mountain honeysucker', long before European naturalists discovered that it visits flowers and possesses a tongue that is well adapted for nectar feeding. It was not until the great period of European exploration beginning in the sixteenth century that written accounts of bird–flower interactions began to appear. Some of these are no doubt based on ideas that the explorers gleaned from encounters with the local people, others were first-hand observations. A selection of these accounts follows, and they should convey something of the awe and wonder European colonists and explorers felt when they finally understood the significance of the birds and their flowers.

Oviedo and others

The earliest published European description of birds that visit flowers is credited to Spanish explorer and writer Gonzalo Fernández de Oviedo y Valdés (1478–1557), who first visited Hispaniola in 1514.

Upon his return in 1523, Oviedo (as he's usually referred to) wrote an initial account of the plants and animals that he had encountered in *La Natural Hystoria de las Indias*, published in 1526. Here he is describing the *paxaro mosquito*, literally the 'mosquito bird':

> There are some little birds so small that they are no larger than the end of the thumb … it is so swift in flight that it is as impossible to see its wings as it is those of a beetle or a bumblebee. There is no person who sees it fly but that thinks it is a big bee.[4] They build very small nests … They are so small that they look like the little birds placed by illuminators in the margins of books of hours. Their plumage is very beautiful; they are gold, green, and of other colours. Its bill is as long as its body, and as slender as a pin. They are very bold, and if a man climbs a tree where they are nesting, the bird attacks his eyes and darts away and returns to the attack with incredible speed.

The first account of a hummingbird in English is by Roger Barlow, who was in South America with Sebastian Cabot in the 1520s. Barlow's manuscript *A Brief Summe of Geographie* was presented to Henry VIII, though not published until 1932. In it he writes of

> smal byrdes which be no bigger of bodie then the toppe of a mans thombe but thei have the goodliest colored fethers that ever man might se, the colours wold chaunge in moving of them as it were chaungeable silke. We toke one of them alive and kept it in a cage and was verie tame … but it lived not long for lacke of knolege to diet it or other keping.

4 Stoudemire translates this as 'bumblebee', but this seems to be closer to the meaning. Thanks to Mary Price for clarification.

Perhaps if they had asked the locals about its diet this bird might have lived longer.

Almost two centuries later, in 1693, British botanist Nehemiah Grew was writing to the Royal Society about the confusion concerning how hummingbirds fed:

> You see it is believed he feeds on some Juice he sucks off, or out of Flowers … Whether may not this Bird rather feed on small Insects whereon many Birds feed, some whereof lie in the bottom of most Flowers, and for which, this Bird hath a Bill? Whereas a Bee that sucks hath a Siphon or hollow Probe.[5]

As we'll see, the debate about birds feeding on nectar versus insects was to rumble on for a further two and a half centuries.

Another early description is provided by the French explorer-priest André Thevet (1516–1590), who in 1555 accompanied a fleet of ships sailing to 'France Antarctique', a colony near what is now Rio de Janeiro in Brazil. He remained there for 10 weeks, and on his return he wrote *Les Singularitez de la France Antarctique*, published in 1557. An English translation appeared in 1568 as *The New Found Worlde, or Antarctike, wherein is contained wōderful and strange things, as well of humaine creatures, as Beastes, Fishes, Foules, and Serpents, Trees, Plants, Mines of Golde and Siluer*. One of the 'Foules' that Thevet describes was called a *Gouanbuch* by the locals, and his description is clearly that of a hummingbird:

> I will not yet forget another bird named Gouanbuch, the which is no bigger than a great Flie, the which for all that it is little is so faire

5 This was a follow-up to a short piece that Grew had published on behalf of a Mr Hamersly from Barbados. Writing on the American Ornithological Society blog, Bob Montgomerie suggests that this was the first ever paper devoted just to birds published in a scientific journal: https://americanornithology.org/bird-paper-one.

> to see too, that it is impossible to see a fairer: his bill is somewhat long and slender, and his colour grayishe, and although to my judgement it is the leaste birde living under the skye, neverthelesse it singeth very well, and pleasant to heare.

It's a confusing description, as no hummingbird, to our ears at least, 'singeth very well', and 'so faire to see' hardly equates to 'grayishe' colouration. Not all of what he was writing was based on his own experiences, and Thevet has been considered an unreliable source for some of what he describes. Nonetheless there aren't many birds in Brazil that could be described as 'no bigger than a great Flie'.

None of these earlier sixteenth-century accounts mentions hummingbirds visiting flowers, though it's shown and discussed in the later *Florentine Codex* (originally entitled *La Historia General de las Cosas de Nueva España*, 'The universal history of the things of New Spain'). This manuscript was written by the Spanish monk Bernardino de Sahagún between 1545 and 1590, and it contains two illustrations of hummingbirds – in one of which a bird is probing a flower. Although the *Codex* is attributed to Sahagún, he used indigenous Nahua scholars and artists for both the accounts and the illustrations. These local people were presumably more familiar with hummingbirds and their natural histories. Writing in 1994, Italian researcher Fernando Ortiz-Crespo pointed out that at least some of the text is an accurate depiction of hummingbird ecology, including descriptions of their flight, the fact that they usually lay two eggs, their torpor in cold weather, and that they feed on 'flower dew'.

Catesby in the Caribbean

English explorer Mark Catesby (1683–1749) made a couple of trips across the Atlantic in the first quarter of the eighteenth century. He wrote up his observations in *The Natural History of Carolina, Florida and the Bahama Islands*, issued in 11 parts between 1729 and 1747. Each part contained 11 coloured plates, and the whole book represents an early and detailed account of North American and Caribbean plants and animals. Of the birds that Catesby included in his book, the now-extinct (or possibly still extant – the debate continues) Ivory-billed Woodpecker gains the most attention from ornithologists. However, his descriptions of hummingbirds are also of special interest. In a section called *Mellivora avis carolinensis* ('the honey-eating bird of Carolina'), Catesby describes 'the humming-bird' thus:

> There is but one Kind of this Bird in Carolina, which in the Summer frequents the Northern Continent as far as New England. The Body is about the Size of a Humble Bee. The Bill is strait, black, and three Quarters of an Inch long. The Eyes are black; the Upper-part of the Body and Head of a shining Green; the whole Throat adorned with Feathers placed like the Scales of Fish, of a crimson metallic Resplendency; the Belly dusky white; the Wings of a singular Shape, not unlike the Blade of a Turkish Cymiter ; the Tail Copper-Colour, except the uppermost Feather, which is green. The Legs are very short and black. It receives its Food from Flowers, after the Manner of Bees; its Tongue being a Tube, thro' which it sucks the Honey from 'em. It so poises it self by the quick hovering of its Wings, that it seems with-out Motion in the Air. They rove from Flower to Flower, on which they wholly subsist. I never observed nor heard, that they feed on any Insect or other thing than Flowers. They breed in Carolina, and retire at the Approach of Winter.
>
> What Lerius and Thevet say of their Singing, is just as true as what is said of the Harmony of Swans; for they have no other

> Note than Screep, Screep, as Margravius truly observes.[6]
>
> Hernandes bespeaks the Credit of his Readers by saying, 'tis no idle Tale when he affirms the Manner of their lying torpid, or sleeping, all Winter; in Hispaniola and many other Places between the Tropics I have seen these Birds all the Year round there being a perpetual Succession of Flowers for them to subsist on.

In these passages Catesby provides information that reflects some of the science that we've already discussed in this book, including flower feeding, migration, torpor, and the importance of flowering phenology.

In many of these early observations by Europeans, the comparisons with insects such as hornets, mosquitoes, beetles and bees are understandable, given the biogeographic disparity that I set out in the previous chapter. In the eighteenth century, Europeans were still trying to make sense of an entirely novel bird behaviour: no bird that they had seen before flew in the fashion of a hummingbird. They therefore referred to familiar insects that provided analogies for their movements and, in some cases, their flower visiting.

Rumphius in the Old World

The first mention by a European of birds other than hummingbirds visiting flowers that I can locate comes from an early German explorer, Georg Eberhard Rumphius (1627–1702). He worked as a botanist for the Dutch East India Company in Indonesia and observed and wrote about local natural history in the 1660s and

---

6 Margravius refers to German naturalist Georg Marcgrave or Marcgraf. 'Lerius' is presumably Jean de Léry, who, in his *Histoire d'un voyage fait en la Terre du Bresil, autrement dite Amerique* ('History of a voyage to the Land of Brazil, also called America', 1578), describes the Gonambuch (Thevet's 'Gouanbuch') as 'a singular marvel and a masterpiece of smallness'.

1670s, though his work *Herbarium Amboinense* did not appear until the 1740s. In volume five of that publication he discusses the natural history of some tropical mistletoes and describes seeing a small bird similar to a species of kinglet such as a Goldcrest, that 'has a long curved bill, by which it sucks nectar from several flowers'. That translation was provided by the Dutch botanist and entomologist Willem Marius Docters van Leeuwen in 1954, who goes on to say that the only bird in the region where Rumphius was working which resembles a kinglet is the Ashy Flowerpecker. This bird is known to feed on mistletoe nectar, pollen and fruit. Many of the flowerpeckers are associated with these mistletoes in that region, acting as both pollinators and seed dispersers: Rumphius was describing 'double mutualisms' centuries before the term was coined.

Despite Rumphius's observations, over two centuries later the influential Indian ornithologist Sálim Ali still felt the need to push back against scepticism that birds fed on nectar, disbeliefs which we can trace directly back to Nehemiah Grew. In 1932, Ali wrote:

> There has been a tendency to attribute the object of birds' visits to flowers solely or chiefly to a search after the attending insects, and the importance of the honey has been unwarrantably belittled. As I have pointed out elsewhere … there seems no reason for doubting that the visits of specialized birds to the flowers of their choice may be exclusively in quest of the nectar. As mentioned before, flower-nectar is rich in carbohydrates and provides excellent nutriment.

This idea, that birds visit flowers mainly or solely to collect insects, was prevalent in the nineteenth century. In one of his contributions to the multi-volume *Zoology of the Voyage of H.M.S. Beagle*, Darwin described his observations of several species, including the

Giant Hummingbird – and he echoed the prevalent nectar scepticism in noting:

> Although flying from flower to flower in search of food, its stomach generally contained abundant remains of insects, which, I suspect, are much more the object of its search than honey is.

Ali recognised that this misunderstanding was, in part, due to naturalists with disparate interests interpreting the evidence in different ways, and made a plea for greater cooperation between those who study birds and those who study plants and their flowers:

> While it is but rarely that an ornithologist also possesses sufficient competence in botany to be able to conduct research of this nature without the aid of a specialist, and vice versa, an intimate co-operation between the two is clearly indicated for obtaining optimum results.

On to Australia and the Pacific islands

The oldest depictions of flower-visiting birds currently documented are thought to be the stencilled images at an Aboriginal rock art complex in Arnhem Land in northern Australia, known in the local Maung language as Djulirri. These beautifully rendered images were made at least 9,000 years ago, by painting around an actual, and presumably dead, honeyeater.

Millennia later, one of the first Australian birds to be scientifically described was the New Holland Honeyeater, by British naturalist John Latham (1740–1837) in his *Index Ornithologicus* of 1790. As was typical at the time, the description was purely of the bird, with no indication of its natural history. It was a little later, in 1825, that Meliphagidae (literally 'honey eater') was coined as the scientific name for the family by Irishman Nicholas Aylward Vigors (1785–1840). In the same publication, Vigors compared the flower-feeding habits of hummingbirds, honeyeaters and sunbirds. The contemporary taxonomy of these three families was rather confused, as part of the sunbird family was classified separately as 'Cinnyridae' and grouped with the hummingbirds as birds that feed from flowers 'entirely on the wing', whilst the 'Nectariniadae' [*sic*] 'hop from flower to flower'. Vigors never saw the birds in the wild, and was reliant on preserved specimens and colonial accounts. Half a century later, *A Monograph of the Nectariniidae, or Family of Sun-Birds* by George Ernest Shelley (1840–1910) provided copious first- and second-hand accounts of these birds, as well as clarifying the taxonomy.

At the same time as Europeans and colonial Westerners were absorbing all this knowledge about the birds that feed on flowers, they were also becoming smitten with their colour and flamboyance. We no longer wear stuffed hummingbirds on our hats or display them gaudily in vitrines in our homes – these are now the domain of museum collections. Nonetheless, those flower-dependent birds have captured the imaginations of marketing people, artists and the public, the world over.

Guitars and songs, wine and cheese

Hummingbirds, it seems, turn up almost anywhere as emblems or reference points. The number of musical associations of hummingbirds, for example, is staggering, in spite of their inability to sing beautifully. A search on the lyrics.com website revealed almost 900 recorded songs that include the word, often to symbolise love. In addition, seven recording artists have hummingbird in their name, and 11 albums use it in their titles. Everyone from Frankie Lane and B. B. King to Cat Stevens and Jimmy Page has a song called 'Hummingbird'.

'Sunbird' fares more poorly, just 24 lyrics and one album, while 'honeyeater' lags behind with a miserly three lyrics to its name.

The musical associations of flower-visiting birds are not limited to songs, however. In 1960 the Gibson Guitar Corporation introduced a new acoustic model named the Hummingbird. The pickguard featured the bird visiting a flower, a rare case of a brand highlighting the ecological interaction rather than just the bird (Plate 20), and some later custom models also featured a hummingbird on the headstock. The plant on the pickguard is a Trumpet Vine, native to eastern North America, which fits with the original location of Gibson's factory in Kalamazoo, Michigan. The pickguard art was created by luthier Hartford Snider, who was a keen naturalist as well as a builder of instruments. The guitar was well received by professional musicians, and famous players include Keith Richards and Brian Jones of the Rolling Stones, and Thom Yorke of Radiohead.[7]

This is not the only guitar to feature a hummingbird. Paul Reed Smith (PRS) Guitars have traditionally used different species of birds inlaid into the neck as fret markers on their instruments. A homage to Smith's mother, who was a keen birdwatcher, the inlays always appear in the same order, with a Ruby-throated Hummingbird at the seventh fret (Plate 21). As far as I know, 'Sunbird' and 'Honeyeater' have never been used to brand guitars, and ecologist Richard Hobbs, whose excellent blog *The Nature of Music* (www.the-nature-of-music.com) explores the relationships between guitars and conservation, has never come across one either.

Sunbirds and honeyeaters have featured on coins and bank notes, however.[8] An Orange-breasted Sunbird is depicted on a ten-rand coin that was issued in 2016 to commemorate the UNESCO Kogelberg Biosphere Reserve in South Africa, and there's an Eastern Spinebill on the Australian five-dollar note. Hummingbirds visiting flowers have also featured on coins and notes from Brazil, Suriname, and Trinidad and Tobago, amongst other places.

For many years the British ten-pound note featured a portrait of Charles Darwin and a prominent hummingbird visiting a flower. It was criticised by British biologist Steve Jones because, as he rightly pointed out, there are no hummingbirds on the Galápagos.

7 There's an amazing video of Thom playing a song called 'The Clock' using his vintage Hummingbird. The guitar has been played so often that the bird has been worn off the pickguard: www.youtube.com/watch?v=Z1nFB-R-_gI

8 I could also mention stamps, but there's only so much room in a book.

But there's more to Darwin than just those islands, and he certainly encountered hummers in South America.

Hummingbirds are so familiar and iconic that they have been used commercially to brand a wide range of goods and services. I was stumped, however, when browsing the refrigerated cabinets in a supermarket in Denmark I spotted one on the packaging of a circular French cheese named *Kolibrie*. It took me a moment to get the pun. Sunbirds and honeyeaters, it seems, are less widely used, but they do feature on brands from South Africa and Australia, respectively. Hummingbirds occasionally feature on the labels of Californian wine bottles (Plate 22) and also beer (Plate 23). But – pushing the phylogenetic boundaries – I was especially pleased to spot a South African wine label featuring a Cape Sugarbird perched on the stem of a pincushion inflorescence (Plate 24).

Our current understanding of the ecological and evolutionary importance of birds that pollinate flowers, and flowers that are visited by birds, is built on both indigenous knowledge and colonial observations. In a similar way, traditional cultural and symbolic values have resonated with everyone from guitar builders to cheese makers. The cultural significance of nature, and how this benefits humanity, is one of the four main categories of what are termed 'ecosystem services', the ways in which nature supports human populations. When it comes to flower visitors, however, the main service they provide to humans is as pollinators of crops and edible wild plants. Although bees are the most important group in this regard, birds also make a contribution, as we'll see in the next chapter.

CHAPTER 15

Feathers and fruits

Jetlagged but eager to start exploring Brazil, I resisted the urge to collapse into bed. Instead, I took a walk around the landscaped grounds of the University of Campinas, where I was to be based for a week teaching a postgraduate course in pollination ecology. The campus was beautifully laid out, with both formal planting and patches of remnant native vegetation. I quickly clocked up a long list of birds that I couldn't identify and would have to check later using the guide I'd brought with me. Close to the campus hotel I spotted vivid red flowers on a large shrub, suggesting to me that it could be hummingbird-pollinated. Examining the shrub up close I realised that it was actually a pomegranate, a native of Asia. As I watched, however, a male hummingbird approached and took nectar from several flowers in succession. Slowly I backed away from the pomegranate and positioned myself on a convenient bench, all the better to observe this encounter between a New World bird and Old World flowers.

I don't know if that hummingbird was pollinating the pomegranate; there were unripe fruit on the plant, but that could have been the result of visits by native Brazilian bees and introduced Western Honey Bees. In European and Asian pomegranate orchards, bees are the main pollinators despite the fact that the large red, bell-shaped flowers have all the appearance of being bird-pollinated (as we saw in Chapter 8, it's a myth that red flowers are invisible to insects). However, I have seen video footage of a Purple-rumped Sunbird taking nectar from pomegranate flowers, and there's a published report of both legitimate flower visits (in the sense of entering the flowers from the front) and nectar robbing by Purple Sunbirds. Both of these accounts are from India, where pomegranates are (arguably) a native species, albeit cultivated, a point I'll come back to later. The authors of the latter study, P.D. Kamala Jayanthi and colleagues, did not assess the contribution of visits by these sunbirds to pollination but did find that robbed flowers that had holes incised in the base of the corolla usually fell off. In addition, the birds' 'swift flight movements and fluttering of wings within the plant canopy also caused direct damage to flowers/buds making them fall'. The researchers go on to suggest planting alternative nectar sources for the sunbirds around pomegranate plantations where they are a problem, in order to draw them away from the crop. If only all 'pest' control were as benign.

Other than these observations, the role of sunbirds and other avian flower visitors as pollinators has never really been studied in pomegranates, though the Chinese painting in Plate 19 suggests that generalist warblers may be pollinating visitors. The problem is that pomegranate is an ancient crop that for millennia has been cultivated and artificially selected for particular characteristics, such as colour and taste of the fruit. At the same time, flower traits have also changed, so who knows what its ancestral pollinators were? In addition, as both an ornamental and a crop plant pomegranate has been spread around the world and its flowers now encounter nectar takers, such as hummingbirds and New World bees, that would never have been part of its ecology before people took an interest in the fruit.

Pomegranate in its small way is a microcosm of a vastly bigger set of issues and questions. Human populations across the planet are locked into complex economic, agricultural, social, political and ecological systems that provide them with food or, in some circumstances, fail to provide that sustenance. The complexity of these food systems requires that they are studied from a multidisciplinary perspective, because no one subject, or individual, can possibly do justice to this crucial topic. Understanding how food arrives on our plates, and ensuring that it's sustainable into the future, therefore requires economists, agronomists, sociologists, political scientists and ecologists to work together.

One of the most pressing questions related to food systems is how we ensure that our food supplies are resilient to the current and future challenges of war, pandemics, climate change, economic shocks, biodiversity loss and a host of other factors. Some of these are predictable, others are not, except that recent and distant history tells us that such challenges are always going to be a feature of our societies and we need to prepare as best we can.

When it comes to biodiversity loss, a great deal of attention has been paid to the pollinators of plants for human consumption. It's estimated that about three-quarters of the world's most important crops are animal-pollinated to some degree, mainly by insects, so we have to add entomologists into the expert mix I described above. The value of pollinators is not only that they increase crop production. Seeds, nuts and fruits that are animal-pollinated often contain greater amounts of essential vitamins, minerals and antioxidants, compared to wind-pollinated cereals or tuber crops such as potatoes. In short, pollinators enhance the health of human populations – of that there is no doubt.

Of those pollinating insects, it's the bees that get most of the attention. From the warm dry almond monocultures of California, to the cool, damp apple orchards of western England, and the small-scale coffee producers in the tropical Global South, bees are seen as being pre-eminent. This includes both managed insects, such as various species of honey bees that are often integrated into crop production to diversify a farmer's income stream, and wild populations of social and solitary species. When I reviewed the topic of agricultural pollination for my previous book, drawing heavily on Alexandra-Maria Klein's widely cited 2007 paper, 'Importance of pollinators in changing landscapes for world crops', plus more recent studies, I found that honey bees (of all types, not just the Western) have been recorded as pollinators of about 40% of the major animal-pollinated crops. Usually they are not the only pollinators, and they act in concert with bees, hoverflies and others.

Bees, however, certainly don't have a monopoly on agricultural pollination. Take the multi-billion-dollar cacao industry, for instance, which provides us with chocolate in its many forms. The production of cacao beans is mainly the result of flower pollination by tiny flies just a few millimetres in length. Reviewing these other pollinators, Romina Rader and colleagues found that, depending on the crop and the location, between 25% and 50% of pollinating flower visits were performed by insects other than bees.

What about the birds? I'd love to be able to tell you that birds are important as pollinators of agricultural food crops, but the truth is that they are not, at least compared to insects. The various reviews and studies of agricultural pollination show that on a global scale birds play a relatively minor role in pollinating our food, and there are no bird-pollinated food plants that are widely exported. Bird pollination of food plants, however, is certainly not negligible at the more human level of local food production by small-scale, often subsistence farmers, especially in the tropics. Perhaps even more importantly, they pollinate wild plants that are collected by people to supplement their farming or are even a major component of the traditional sustenance of some hunter-gatherer groups. And as we'll see, bird pollination of non-food crops is not insignificant.

Many of these tropical fruits deserve to be more widely known and appreciated. One such fruit is the guaba, which I'll discuss in Chapter 19, and another is a plant that goes by a bewildering variety of monikers.

Feijoa by any other name

Depending on who you talk to, be it a farmer, a consumer or a taxonomist, this tropical crop, which I first encountered in Brazil, may be known as 'feijoa' (pronounced *fay-zhow-uh*), 'Brazilian guava', 'pineapple guava', 'fig guava' or 'guavasteen'. It's not a true guava, despite the name, though it is distantly related as a member of the same plant family, the myrtles and eucalyptuses (Myrtaceae). Its scientific designation is *Feijoa sellowiana*, but to complicate matters taxonomists sometimes place it in a different genus and refer to it as *Acca sellowiana*. Whatever we call it, the plant is native to South America, from where it's been domesticated and spread around the world as a crop suitable for warm-temperate or subtropical conditions. I even planted it in the garden of our old house in Northampton, England, where it grew well – though it rarely flowered and never set fruit.

That fruit is delicious, sweet and aromatic with hints of pineapple – it has a flavour all of its own. It's rarely found in European shops, as the fruit is easily bruised and has a short shelf life, so is difficult to export and market outside of the main areas of cultivation.

If we step back from the fruit stage for a moment and consider the flowers, we see that they have all the hallmarks of being hummingbird specialists. They are quite large, with pinky-red and white petals surrounding a vivid crimson cluster of stamens, each tipped with an anther containing bright yellow pollen, as if someone had used the bloom as a brush to paint a beach scene. In the centre lie the female parts of these hermaphroditic flowers, the stigma, style and ovary, which turns into the fruit following pollination and fertilisation.

There is, however, no nectar in the flowers – which means that any hummingbirds that visit them quickly learn to avoid these blooms. Despite that, a review of feijoa pollination biology by Fernando Ramíreza and Jose Kallarackal concluded that birds are important pollinators of the flowers, at least in its native range. But these birds are not looking for nectar, or even pollen: they eat the petals.[1]

The pollinators are mainly birds from families such as tanagers, thrushes and mockingbirds, all of which are known to regularly consume fruit, plus some like the Red-legged Honeycreeper that also feed on nectar. It's the fruit-feeding behaviour of these birds that is the key, however, because feijoa is one of a handful of plants that offer sugar-rich petals as a reward to normally frugivorous birds. This type of pollinator reward strategy, in which the boundary between pollination and seed dispersal becomes blurred, occurs in a few additional members of the same plant family, and has rarely been documented in other families (see Chapter 7).

Experiments within the natural range of feijoa in southern Brazil have shown that excluding birds by netting the shrubs, but still allowing insects access, reduces the proportion of flowers that

---

1 For humans as well as birds, the petals of feijoa are perfectly edible, and have been described as 'spicy' to the palate.

turn into fruit from 47% to 31%. Most cultivars of feijoa are self-incompatible and cannot produce seeds and fruit when fertilised with their own pollen. Insects therefore clearly have a role to play in feijoa reproduction, and the pollination system of this crop is best described as being mixed bird–insect generalist. In New Zealand, far beyond its natural range, feijoa is effectively pollinated by two non-native birds (Eurasian Blackbird and Common Myna) and one that is a native of those islands, the Silvereye. However, studies in other parts of the world have shown only insects to be the pollen vectors. Just to add to this mix, in Colombia, ecologists Carlos Matallana-Puerto and João Custódio Fernandes Cardoso recently documented Brown Rats, as well as birds, as feijoa pollinators in the amusingly titled article 'Ratatouille of flowers! Rats as potential pollinators of a petal-rewarding plant in the urban area'.

To summarise, then, birds are not the only pollinators of this locally important crop, but they certainly play an important role in its unusual pollination system. Things are a bit less clear-cut, however, for the next fruit.

Going bananas

Kit Prendergast, an Australian pollination ecologist, posted on social media in 2022 a video showing the flowers of a banana growing in her garden in Queensland that was being visited by a pair of Blue-faced Honeyeaters. Above the flowers hangs a large bunch of ripening fruits, suggesting that something, perhaps these birds, is effectively pollinating them. I commented on Kit's video that I had read a lot of statements claiming that bananas are bird-pollinated, but that it was unclear to me whether there's much in the way of data to support this claim.

So often in science, especially in ecology, an observation or a comment results in a search through the older scientific literature to see whether statements hold true. This can result in whole days wasted trying to get to the bottom of vague and elusive claims, as I found with banana biology. I'll not go into the intricacies of modern banana hybridisation and domestication (the details of which are still being worked out by geneticists) or why most of the bananas that we buy in shops in the global north are a single, seedless variety that does not require pollinators to set fruit. Suffice to say that many of the wild or near-wild bananas that are grown in the tropics and mainly used for local consumption are, at least in part, bird-pollinated in the natural range of these plants, which is tropical Africa and Asia. Sunbirds are thought to pollinate bananas in Nigeria, while across Indonesia, banana species have been shown to be pollinated by birds such as the Copper-throated, Scarlet and Javan Sunbirds, as well as nectar-feeding bats. In China, the Yunnan Banana is also pollinated by bats and by Little Spiderhunters, which incidentally are also thought to be one of the pollinators of Asia's infamous durian fruit.

There's a longstanding suggestion that those banana species that have flowers which hang pendulously are bat-pollinated, whilst the species with upright flowers utilise birds. However, Kit's video and the published data on mixed bat + bird pollination in some species would seem to refute that idea. As we often find when we try to use flower traits to predict pollinators, there are exceptions to the rules.

Although birds are not frequent pollinators of commercial food crops, it's clear that they are often involved in the reproduction of wild-growing relatives of the more familiar crops. The bananas are one example, but there are similar stories to be told about papayas and pineapples. For example, Peter Kwapong and Andreas Kudom in Ghana found that pineapple fields attracted large numbers of flower visitors, especially butterflies, but also four species of sunbirds. None of these were effective pollinators, as the pineapple fruit is produced without the requirement for pollination. However, their work does hint at the ancestral condition of this familiar fruit, which would have been hummingbird-pollinated in its native South America. Looking at it from the flower-visiting bird's perspective, of course, crops that provide nectar can be an important supplement to wild sources, even if the farmer does not benefit.

All of this is important not necessarily for immediate commercial considerations, but because such plants are a vital food source for huge swathes of the world's human population, especially in tropical and subtropical rural areas in Africa, Asia and South

America. In the future we will need these wild genetic resources to augment and bolster the widely grown crops that are often genetically un-diverse and at risk from emerging plant diseases and the impacts of climate change. A good example is Golden Camellia, the tea from which is drunk for its supposed health benefits. This winter-flowering plant is rare in its native southern China and Vietnam, where recent work by Shi-Guo Sun, Shuang-Quan Huang and colleagues has shown that the Crimson Sunbird is an important pollinator. Some of those researchers have worked with another winter-flowering tree, loquat. A member of the rose family, like so many orchard crops, loquat has been cultivated for over 1,000 years for its delicious fruit. Like Golden Camellia, it is bird-pollinated in China, though this time by generalist passerines including Light-vented Bulbul and Warbling White-eye.

All of the world's crops have indigenous roots, all were grown in their native range by local peoples, and all have been gradually transformed by selecting desirable traits. The wider dispersal of these crops, often globally, has happened to a rather small proportion of the fruits, vegetables and cereals that have been selectively bred in this way. The potential therefore exists to develop some of the many, many edible fruits and seeds that are collected and consumed by indigenous peoples, but which are almost unknown to modern agriculture. Some of these are bird-pollinated – for example the fruit of the Canary Island Bellflower, or Bicácaro, that we encountered in Chapter 12. This was almost certainly exploited by the first peoples of the Canary Islands – the Guanches – and the tradition of collecting and eating it exists to this day. I've tried it and the taste is not exceptional, but with appropriate horticultural selection it could be a potential future crop. There are many other such examples from all over the world. In Argentina, for example, Pablo Gorostiague tells me that people collect and consume the fruits of a few wild hummingbird-pollinated cactus species, though in general 'most bird-pollinated cacti produce quite small fruits that are not eaten'.

In the Sonoran Desert of southwestern North America, another, partially bird-pollinated cactus tells a different story. The indigenous Tohono O'odham ('Desert People') have inhabited that challenging landscape for thousands of years, where the harvesting of Saguaro fruit has been an annual event for at least five centuries and is probably much more ancient. Not only are the fruit and seeds of this cactus consumed, but the pulp is made into a sweet syrup that is turned into a fermented drink. The resulting fruit wine is then consumed at a ceremony to mark the start of a Tohono O'odham new year and to bring the summer rains, which stimulate the cacti to flower, and so the cycle continues. The whole practice of collecting and processing Saguaro fruit is filled with cultural and religious significance for the Tohono O'odham and is still considered a 'reaffirmation of their relationship with their traditional environment'.

In a recent review of the crops, and their pollination requirements, grown in agroforestry systems in the Brazilian Amazon region, William Sabino and fellow researchers discovered that birds pollinate 12% of the food plants. In contrast, a wider review of pollinators of the main Brazilian crops found that only about 2% were bird-pollinated. The crop that was most dependent on bird pollinators in both of these reviews was bacuri, a tree in the garcinia family (Clusiaceae) for which a wide diversity of species, especially various tanagers and parrots, were considered essential to the crop. William and colleagues argue that

> Agroforestry systems, which are the intentional integration of trees and shrubs into crop and animal farming systems, are a more sustainable production approach that has been increasing in several forested areas around the globe.

Anyone who has tried to attract birds to their own gardens knows that it's the complexity of the structure of the vegetation, with tree, shrubby and ground-cover layers, that results in the greatest diversity. So perhaps it's not surprising that many birds are associated with these agroforestry systems – which provide an important alternative to modern monoculture approaches to farming.

In addition to the fruit being eaten in Brazil, the oil produced from bacuri seeds is marketed as a moisturiser and health product that can 'promote healthy hair, skin and nail growth and … reduce the appearance of blemishes and scarring'. There are, of course, lots of plants that are grown or collected commercially not for their food, but for the other good things that nature provides to enhance our lives.

Roubik's review: beyond food

In 1995 the Food and Agriculture Organization of the United Nations (FAO) published a book-length report entitled *Pollination of Cultivated Plants in the Tropics*. Edited by American bee expert David Roubik, it summarised what was then known about agricultural pollinators outside of the temperate north. Although it's been superseded by later publications, it's still available for free download from the FAO's website and is a useful summary of the state of knowledge at that time. Appendix 1 of that report is especially interesting because it provides a summary of the plants that are 'grown in tropical countries and harvested in semi-wild habitats … [which are] … widely accepted as economically important'. In any such compilation, a line has to be drawn as to what to include and what to leave out; in this case, not included are 'species that are sold as ornamental plants, used for ordinary timber or firewood, or used for home remedies'. Nonetheless, there are 1,330 entries on that list, and for many of them there's an indication of the observed or predicted pollinators.

Surprisingly, given what I said above about how rare bird pollination is in crops, more than 50 of the 1,330 plants listed have birds as the confirmed or suspected pollinators. Why this discrepancy? There are several reasons. One is that in almost all of the cases, birds are just one of several types of pollinators – the crops are quite generalist in their pollination systems and bees, bats and other animals can do the job as well as birds. The assumption amongst agricultural pollination researchers is often that birds are less effective pollinators of crops, which may be true, but needs to be tested (ironically, feijoa is listed simply as bee-pollinated).

More significantly, the FAO list contains many non-food crops that are at least partially bird-pollinated. Examples include various species of acacia and eucalyptus which are cultivated for their gums, essences or dyes, as well as fibre plants such as kapok, silk-cotton and New Zealand flax. In the current 'one in three mouthfuls' hyperbole compelling us to conserve pollinators because of their role in food security and the provision of nutritious (as opposed to starch-filled) crops, we often forget that there's an important role for the other plants that enhance our existence, ensure our survival, or just make life more comfortable, such as by providing fibres for clothing and other material goods. It is important that we recognise that birds also play a far wider role in agriculture than just as pollinators.

Birds and ecosystem services

The role of birds as pollinators of crops or plants that we otherwise exploit falls under the general label of 'ecosystem services', otherwise known as 'nature's contributions to people', amongst other terms.[2] These are the materials and processes that nature provides for us as part of a fully functioning biosphere, including things like fresh water, production of soil, decomposition of organic materials, and of course pollination of crops. As we've seen, birds are important as pollinators of some crops, though they can also be responsible for 'ecosystem *disservices*' by eating the buds of fruit crops, for instance. In agriculture, this is more than balanced, however, by an ecosystem service in which birds excel – pest control.

In a study that generated a lot of newspaper headlines back in 2018, researchers Martin Nyffeler, Çağan Şekercioğlu and Christopher Whelan estimated that every year the world's arthropod-eating birds devour between 400 and 500 million tonnes of flies, beetles, moths, spiders and other arthropods. Their study assessed all kinds of ecosystems, but if we look just at agricultural

2 This concept, along with 'natural capital', is not appreciated by all environmentalists, on the grounds that we should value nature for its own sake rather than putting a human value on it, monetary or otherwise. Regardless of this, we do benefit from nature, and some of that benefit is economic. That doesn't mean that ecosystem services provide a mechanism for selling off nature to the highest bidder.

land, the statistics are just as striking: in and around crops, birds consume about 28 million tonnes of these invertebrates per year. Not all of these are pests, by any means,[3] but some are, and birds have been shown to successfully control highly damaging insects such as the Coffee Berry Borer.

Granivorous birds, likewise, eat a lot of weed seeds, though not as many as rodents, it has to be said. Much more important for plants that are harvested from the wild, or which are semi-domesticated, is seed dispersal by birds. In the Mariana Islands, for instance, donne' sali chilli is a variety of the Wild Chilli that has both cultural and economic significance to the peoples of that Pacific Ocean archipelago. It's collected from the wild and sold locally, where it's used in traditional foods. As with all chillis, the fruits are consumed by birds and the seeds dispersed in their droppings. However, especially on Guam, the extirpation of fruit-eating birds, thanks to the invasive Brown Tree Snake, has resulted in a decline in the abundance of the chilli. This is an extreme case, because it's occurring on oceanic islands where, as we saw previously, isolation and high numbers of endemic species make them especially vulnerable to human impacts. But there are plenty of other culturally important bird-dispersed plants that also have an economic value – for example European Mistletoe and Holly.

Feijoa, as I noted above, has a pollination system that bridges the gap between rewards for pollinators and rewards for seed dispersers. For the birds, the flowers are functioning more like a fruit than a bloom. A couple of the studies of bird pollination in that plant have been conducted in urban areas – and that's where we are heading next: to the towns and cities of the world and the birds and flowers which inhabit them.

---

3 In fact some will be pollinators – tits are very fond of disemboweling bumblebees and removing their 'honey stomachs', for instance. But hey, a bird's got to eat.

CHAPTER 16

Urban flowers for urban birds

It was still early in the morning, but the heat and humidity of Shenzhen was rising, and locals walking to work had unfurled their sun-shading umbrellas. On my way to the enormous conference centre I was stopped short by the sight of a small yellow bird lying still on the pavement. Crouching down, I carefully picked it up and examined the fresh corpse. It was an adult Swinhoe's White-eye, the pale eye-ring that gives this group its name now closed. I had often seen them flying around the urban forest of Shenzhen, the sixth-largest city in China, but this was my closest encounter. Cause of death was unclear, there was no obvious damage. Perhaps it had flown into a plate glass window on one of the many high-rise office blocks in this rapidly expanding metropolis. I placed the bird on the damp soil of a nearby planter filled with exotic shrubs. No doubt a larger, scavenging bird, perhaps one of the corvids, would come along and take advantage of the free meal.

A city without trees is like a bird without feathers...

It's always exciting for a northern European like myself to visit cities in the world's tropical zones, and compare the way the roads are laid out, the architecture and, especially for an ecologist, the wildlife that inhabits these urban landscapes. Across the tropics, large conurbations vary hugely in this regard. Shenzhen is sleek and modern and rather manicured. In contrast, older cities such as Hong Kong, the first tropical city I ever explored, back in 1993, wear their age like a coat, embroidered with layers of human history and natural colonisation.

The Brazilian city of Campinas, which I visited during my teaching trip in 2013, is also older and more settled in itself, big and busy, hot and hectic. Temperatures when I was there were in the low thirties Celsius in the open, but as soon as we passed beneath the shade of any of the large trees that lined the streets and small squares, the direct heat from the sun was blocked and the shade made it much more comfortable. Urban ecologists have long recognised that city trees provide multiple ecosystem services. They store carbon in their woody trunks, branches and roots, of course, but they also significantly alter the local microclimate. Nowhere is this more apparent than in subtropical and tropical regions, but you can feel their effects even in a north temperate town, where the presence of trees cools parks and pavements, insulating against high temperatures. In the winter, evergreen species in particular also provide some insulation against the cold.

Not only that, but from a human perspective trees are beautiful, of course, and so a city without trees is indeed like a bird without feathers, because trees, like feathers, are both functional and ornamental.

In Campinas, as in other tropical cities, many of the trees were from families familiar to me, such as the figs and legumes and palms. Some were native to the region, others were not. One of the student projects on the week-long course that I ran at the University of Campinas collected data on the origins of the trees in a local park. Of the 64 tree species identified, 45% naturally occur in that region of Brazil. The remaining 55% were from other parts of Brazil, or from other countries. Nonetheless, most of them provided resources for wildlife, such as nectar for flower visitors or fruit for the local birds.

Just occasionally one sees a bird-pollinated tree planted in a city. The most common in my experience are various banksias in Australia (Plate 4), and the Royal Poinciana (from Madagascar) and the African Tulip Tree in the urban tropics and subtropics elsewhere in the world. I've also occasionally encountered large specimens of Poinsettia: when they are given free rein they are a much more impressive plant than their Christmas cousins. The vivid red bracts that surround the clusters of flowers suggest that they may be hummingbird-pollinated in their native Central America, but as far as I know their pollination ecology has not been studied.

But regardless of whether city plants are naturally bird-pollinated, as we have seen, many flower-feeding birds have flexible diets and can exploit nectar from a variety of sources. This fact alone means that the birds that feed on flowers are more diverse in cities than we might imagine.

... and a city without birds is not a city at all

It's estimated that approximately 20% of the world's birds, over 2,000 species, live within the world's towns and cities. Depending upon the extent of the urban area and where in the world it is located, urban birders can expect to see anywhere between about 24 and almost 370 species in their gardens and parks, and on their daily commute. Most of these will be native to the region, with relatively few introduced from other parts of the world. In 2022, a China-based team of researchers led by Alice Hughes was able to draw some major conclusions about the main factors influencing urban avian diversity by assessing more than 780 million records of over 10,700 bird species, including those from more than 48,000 cities around the globe. The biggest factor, perhaps predictably, was the amount of green space within each city. But more subtle effects also emerged from their analyses, such as the amount of night-time light the cities generated, which had a negative influence. That Alice and colleagues were able to assess such a large sample of cities is testament to the popularity of birdwatching as a hobby, of course, but also that birders are willing to upload their observations to the eBird platform, a worldwide database of bird records that was launched by the Cornell Lab of Ornithology in 2002.

For thousands of years people have altered the physical structure of the world around them, building artificial edifices and adding or removing plants. Animals have learned to adapt to these changes and to exploit them, to the extent that some species (think of House Sparrows or Feral Pigeons) would be lost without the presence of people and their dwellings in a landscape. In any garden or urban park, the diversity and abundance of birds depends on a wide range of factors, but the types of plants are especially important: grass lawns will provide foraging opportunities for some birds, but add in dense shrubs and tall trees to provide additional food niches and opportunities to roost and nest, and the avian variety of an area will dramatically rise.

In some parts of the world, this increased diversity will include the birds that regularly or occasionally visit flowers. If appropriate nectar sources are provided, and there are native or introduced flower-feeding birds in the locality, they will find them and exploit them. In northern Europe this might be Common Chiffchaffs and tits feeding on willows or cherries, but in the tropics a much wider range of birds and plants will of course be involved.

The Swinhoe's White-eye that I encountered in Shenzhen I originally referred to on my blog as a Japanese White-eye, reflecting the accepted taxonomy of the species at that time. That was in 2017, the year that China hosted the International Botanical Congress, a six-yearly event that brings together the world's plant scientists – 7,000 of them that year. I'd applied to the Chinese Academy of Sciences, which funded my trip in return for presenting two lectures at the Congress. A couple of years later, I had to revise my identification of the bird when a group of ornithologists from Singapore, China and Indonesia published a taxonomic revision of these birds entitled 'Molecular evidence suggests radical revision of species limits in the great speciator white-eye genus *Zosterops*'.

To cap it all, the Japanese White-eye is now called the Warbling White-eye: in the words of the late, great Kurt Vonnegut, 'So it goes.' Taxonomy is always a moving target, especially in the bird world. But whatever we call these smart, active little birds there's no doubt that they feel completely at home in the urban setting of a Chinese mega-city. Despite its hi-rise, hi-tech, hi-gloss image, Shenzhen is actually quite green, at least in the sense of being heavily planted with trees and shrubs, some native, many not. Within this complex of concrete, steel, glass, leaves, branches and flowers, Swinhoe's White-eyes thrive and are now one of the most abundant birds in this and other south Asian cities. But what about other parts of the world? Which flower-loving birds do we find there?

Back to Brazil

Exploring the diversity of flower-feeding birds was just one of the aims of research by a group of Brazilian scientists, including Pietro Maruyama, when they studied hummingbird–plant interactions in urban green space in the municipality of Cáceres, in the state of Mato Grosso. They also wanted to understand how these city spaces could be planned and managed in the future to make them more appealing to the birds. During 146 hours of surveys they recorded five species of hummingbirds taking nectar from the flowers of 14 plant species, 11 of which were native and part of the remnant vegetation left behind during urban expansion. By far the most common hummingbird, dominating the interactions with flowers, was the Glittering-throated Emerald, a relatively short-billed species that is known to be quite the generalist when it comes to flowers. It's a bird with a wide distribution in tropical South America, and most populations are sedentary, though one of the Brazilian subspecies migrates along the Atlantic coast as far south as the state of Rio Grande do Sul. This study concluded that two factors stand out when it comes to supporting flower-feeding birds in the city. First of all, as much native vegetation as possible should be maintained; even small, remnant patches have a value. Secondly, the timing of flowering is important, because birds such as these are active all year round and indeed breed for a large part of it. If trees and shrubs are to be planted in green spaces, there needs to be careful consideration of their flowering phenologies to ensure that there are no major gaps within the year when flowers are not available. This is especially important for those bird species that are territorial and defend their patches against incomers.

Pietro was one of the postgraduate researchers whom I met during my time at the University of Campinas, and since then he has developed an impressive research career as an ecologist interested in hummingbird–flower interactions. He has been involved in a number of other studies of urban pollinators in Brazil, including a 2020 synthesis of 86 studies that involved more than 500 plant species and over 330 pollinators. Most of the latter were insects, of course – but just under 10% were birds. These were principally hummingbirds, but they also documented other frequently encountered pollinating bird species, including various tanagers, parrots and the Bananaquit. More surprisingly, the introduced House Sparrow was noted as a visitor to flowers of a native coral tree. There have been a few other records of this ubiquitous urban bird visiting flowers in South America, and also in Gibraltar. We can speculate that it's an example of niche broadening or interaction release of the type we find on islands (see Chapter 12): the sparrow finds itself in a new environment and begins to exploit a resource, in this case nectar, that is not often available to it within its native range.

Working in the Brazilian city of Ilha Solteira, in the northwest of São Paulo state, Paulo Silva and colleagues discovered that a different native coral tree is considered a useful source of nectar for urban birds during the marked dry season between April and September. Once again it was seen to attract a diverse assemblage of 17 species, but this time only five of them were hummingbirds. Based on the birds' behaviours when they visited flowers, just seven species were potential pollinators: one of the hummingbirds, the Shiny Cowbird, Variable Oriole and Red-rumped Cacique, and all three parakeets. The rest were nectar thieves, taking the reward without interacting with the sexual parts of the flowers. I think that this is a nice demonstration of how assumptions about the effectiveness of different birds as pollinators can be challenged by data in the field.

When it comes to documenting urban flower-visiting birds, the Americas have taken the lead, with more studies from there than the rest of the world combined. That reflects the diversity of hummingbirds and the fascination that scientists have with these stunning species, as well as the increasing pace of urbanisation across the continent. Other parts of the world are beginning to catch up, however, and we're seeing more research from Australia, Asia and Africa.

In Cape Town, Anina Coetzee, Phoebe Barnard and Anton Pauw have shown that the numbers of both flowers (especially native species) and artificial sugar-water feeders present in an urban garden are important determinants of the number and diversity of specialist sunbirds and generalist passerines. Artificial nectar feeders are increasingly being used to attract pollinating birds into gardens, but their impact on those birds, and the pollination services that they provide to native plants, is only just beginning to be assessed. However, Anton's research group has also shown that it's the long-billed and arguably more specialised birds that are most affected by urbanisation, while the short-billed species are more forgiving of these artificial environments.

Albeit they are ecologically important, there's just a handful of specialist nectar-feeding birds in and around Cape Town – a low diversity compared with what a Brazilian garden can offer.

Jonas's garden

The most spectacular hummingbird encounter that it has ever been my privilege to experience occurred in a private garden nestled within the remnants of the Atlantic Forest of Brazil. Quietly we closed the doors of the car, and carefully made our way along the winding garden path, weaving between exotic shrubs and small trees. Suddenly the garden opened up and the scene before us stopped us in our tracks. There were hummingbirds everywhere. Zipping, weaving, hovering, manoeuvring, like a drawing of an airshow created by the imagination of a small child, trying to fill the page with as many aircraft as possible. But an illustration that was animated, constantly on the move. The guide to birds of Brazil that I had brought with me said that over 80 hummingbird species occurred in the country. As we'd travelled around visiting sites and collecting data, we typically saw two or three species in a day. In one hour, in that garden, we counted eleven.

The property belongs to a retired teacher named Jonas d'Abronzo. He had moved there 12 years earlier, in 2001, and immediately started attracting hummingbirds to the garden with a home-made sugar-water feeder. It was so successful that he added a second, then a third, and then before he knew it he had 13 of them hanging from trees, and from his porch.[1] Keeping that number of feeders going was hard work, as they needed to be refilled several times a day, using up to 5 kilograms of sugar.

As we sat and chatted with Jonas over a cold drink, we tried to work out which species we were seeing and how many birds there were in total. It was difficult to count them, they were moving so fast, zipping around and squabbling with one another, but we estimated that there were at least 100 individuals. Jonas told us that he'd recorded over 20 species of hummingbirds using the feeders. Obviously the number and diversity of hummers in his garden has been artificially boosted, even if it is located in an amazingly diverse habitat, which is humid lowland rainforest on the border of the Serra do Mar State Park. Different species arrive in his garden at predictable times of the year, responding to seasonal changes in flowers across the region. However, he told us he had noticed that some had become more sedentary over time, constantly hanging around the feeders when in previous years they would have disappeared for months on end. That's perhaps not surprising: the birds had learned that the feeders provided a constant and predictable food source, so why move?

We also discussed a concern that Jonas had, which was that his frequent feeding of the birds might affect the reproduction of some of the local forest plants that rely on them for pollination services. This seemed to us unlikely, and a relatively small issue compared with some of the other activities going on in the area, including clearing land for housing and farming. If anything, by feeding the hummingbirds Jonas was likely to positively affect the local populations of birds, which then disperse out into the wider surrounding habitat. On the other hand, a high density of aggressive species competing for food in the same area may displace less aggressive species.

---

1 I gave a short account of this garden in *Pollinators & Pollination*, accompanied by two images of the feeders that Jonas makes from old plastic water bottles (Figure 9.7).

The following year (2014), some of the postgraduates who had accompanied us on the visit performed an experiment on Jonas's property aimed at testing whether the feeders were influencing plant reproduction. This study was subsequently published in the *Journal of Ornithology* as 'Spatial effects of artificial feeders on hummingbird abundance, floral visitation and pollen deposition'. The artificially high numbers of hummingbirds around the feeders resulted in greater visitation of the birds to hummingbird-dependent species of *Psychotria* (a member of the coffee family). This did not, however, result in less pollination of these plants further out into the forest – there were enough hummingbirds out there to ensure the reproduction of the plant.

It's too early to say whether those findings are generally true, because experiments of this kind have rarely been conducted. In Mexico, María del Coro Arizmendi and co-workers found that the presence of sugar-water feeders led to a reduced seed set in one *Salvia* species, but not in a second, probably because that species was also visited by bees. In South Africa, Monique du Plessis addressed some similar questions in her PhD work in the fynbos vegetation adjacent to suburban Cape Town. She discovered that when sugar-water feeders were available, fewer sunbirds were found in the fynbos, and visits by the birds to the flowers of one species of heather were unaffected whilst visits to a second dropped by 16%. Whether pollination was also reduced in that second heather is unknown.

Feeding birds in gardens always has consequences, both intended and unintended, for the animals and potentially for plants as well. It's important to keep those feeders clean, of course, because there's the potential for spreading diseases. There's also anecdotal evidence that the behaviour of some species changes around feeders, with birds becoming more aggressive. In 2019, in response to concerns about the ecological impacts of supplementary feeding and aggression towards people, Costa Rica went so far as to ban all supplementary feeding of animals, including the use of sugar-water feeders for hummingbirds.

Scientists are now studying sugar-water feeders with more seriousness and employing some state-of-the-art technology to assess how much birds rely on them, such as collecting hummingbird breath and analysing the carbon-isotope composition in the exhaled CO_2. This has shown that, even when sugar-water feeders are available, Broad-tailed and Rufous Hummingbirds will visit both those and any suitable flowers. However, a nine-year data set collected at the Rocky Mountain Biological Laboratory in the 1980s by David Inouye and colleagues showed that in years where there were lots of flowers available, Broad-tails tended to ignore feeders, but visited them more frequently when there were fewer flowers. Perhaps those birds are getting something from the chemically complex nectar of the flowers that is not present in the simpler sugar and water mix of the feeders?

Living for the city

In Israel, Palestine Sunbirds have become much more common in the last couple of decades. They have expanded out of their usual, more restricted habitat to feed on the flowers planted in people's gardens. This is a trend that we have seen with other birds all over the world, including Noisy Miners in Australia (a problematic species that we'll encounter in the next chapter) and Eurasian Collared-Doves in Europe. Some birds, at least, are becoming more urbanised. The same is true of people.

The United Nations estimates that around 57% of the world's human population currently live in towns and cities, and that this will rise to 68% by the middle of the twenty-first century. In terms of the physical footprint, towns and cities equate to less than 5% of the habitable land surface of the planet. The actual environmental footprint, of course, is much greater when we consider the food production and other resources that are required to support this urban population. At the same time, it is becoming increasingly clear that people and wildlife can both survive and thrive in urban settings, but only if *both* are effectively supported. It's often suggested that living in a city disconnects us from nature, but that does not have to be the case, because wildlife is everywhere, if we look and take notice. I've met plenty of people who live in a rural setting who have little knowledge of the animals, plants and other organisms around us, and are less attuned to nature than 'townies'. City planners and inhabitants can do much to enhance urban areas by taking ecologically informed decisions when choosing plants, where to grow them, and how they should be managed.

Several important reviews of urban pollinator diversity and abundance have been published recently, mostly focused on bees and other insects, and it's clear from these that the types of pollinators that exist in urban areas are a filtered subset of those occurring more widely in the region. But then that's true of all habitats, including those that are pristine and natural. There will always be species that cannot tolerate the conditions imposed by climate, competition, geology and other factors. In the words of the late British ecologist John Harper: 'The tropical rainforest is a stressful habitat – for a penguin.'[2]

Wildlife, including flower-visiting birds and the plants on which they depend, can coexist with humans and their lives, and under some circumstances the diversity of pollinators may be at least as great as – if not greater than – the diversity in surrounding, intensively farmed agricultural land. The current biodiversity emergency indicates that we need to consider seriously how human activities are affecting the other species we share the planet with, and disrupting the ecological functions on which we all depend. In the final three chapters I want to explore some of these issues, and their solutions, in detail.

---

2 Harper is supposed to have said this at a British Ecological Society meeting, during a terse exchange with a rival scientist, Phil Grime. Both were highly influential plant ecologists who made important contributions to the field, but I find it odd and somehow sad that neither of them paid much attention to the pollination requirements of plants.

CHAPTER 17

Bad birds and feral flowers

It had been a frustrating few days. My students and I had surveyed several populations of Tree Tobacco across the south of Tenerife. We had looked for potential pollinators, checked the flowers for damage, and had loaded up some of the blooms with fluorescent dye powder. This is an old technique for tracking pollinator movements: if any insects or birds visit the flowers, they pick up the powder and move it to other flowers, providing us with indirect proof that they had been foraging. Studying the ecology of species interactions sometimes requires these forensic approaches. Despite this, we'd seen no evidence that Tree Tobacco ever enjoyed any animal attention, even though the flowers are full of rich, sugary nectar. I returned from that trip disheartened. It was only months later, when I was looking at our results in detail, that I realised what an important insight our findings had been: sometimes zero is the most significant data point.

Tree Tobacco, as the name suggests, is a woody relative of the popular, addictive, and sometimes deadly cash crop that's grown across the world. It's a small tree, usually no more than 6 metres in height, and is native to a relatively restricted area in central South America, where northwestern Argentina meets Bolivia. In this region it is never found in high abundance.

The flowers of Tree Tobacco evolved to be pollinated by hummingbirds. They are usually bright yellow, but occasionally individuals with reddish flowers may be found in those native populations. Following the arrival of Europeans in South America, someone had the bright idea of collecting its seeds and growing it in other parts of the continent. The plant is now found in dry, rather disturbed areas across South and Central America, into Mexico and the southern parts of the United States. But it's only ever the yellow form that you encounter: whoever made that first seed collection did so from only one or a few yellow-flowered plants. Throughout the Americas, wherever there's some disturbed ground such as a dry riverbed, a newly graded dirt road or a construction site, Tree Tobacco flourishes in abundance and hummingbirds find it and pollinate its flowers.

Not content to leave this plant on the continent of its origin, someone else decided that Tree Tobacco would be a nice addition to the gardens of Spain. The Canary Islands were the main staging post for sea voyages to and from the New World, and it's almost certain that this is how it ended up on the other side of the Atlantic, spreading into southern Europe and Africa, then China, and ultimately to Australia, New Zealand and the Pacific islands.

Wherever it finds itself, Tree Tobacco co-opts any of the local avifauna with bills and tongues long enough to reach into the tubular flowers and take the nectar. In Israel, this is the Palestine Sunbird. Malachite, Dusky and Southern Double-collared Sunbirds are pollinators in South Africa, where, against expectations, the birds changed their behaviour and started hovering to feed from these hummingbird-adapted flowers. Honeyeaters have also been seen visiting flowers in Australia, but their role as pollinators is unclear.

In southern Europe and the Canary Islands the plant has no such fortuitous partners, and it must rely on self-pollination to persist. Once I understood that fact, our Tenerife data became an important piece of the jigsaw and I set about bringing together the available published and unpublished data, and their owners, to review what we knew about its reproduction. This overview of Tree Tobacco as an invasive alien species (IAS) was published in the *Journal of Pollination Ecology* in 2012, and at the time this was one of the few studies that compared the pollinator requirements of an invasive plant in its native and naturalised ranges.

Make no mistake, Tree Tobacco is a problem, which is why I started to study it with my students on Tenerife. Individual trees can produce more than a million tiny, easily dispersed seeds in a season, and where it becomes established, Tree Tobacco forms dense stands that exclude other plants. It is also poisonous to livestock and wild animals, and is listed as a worldwide problem in the *Global Invasive Species Database*. The plant is regionally considered a conservation threat in areas as diverse as Hawaii, Europe and South Africa. Tree Tobacco is a particular concern in those parts of the world that are becoming drier due to climate change, and the coming decades might find this bird-specialist popping up more widely across the world.

Tree Tobacco is just one of hundreds of plants that are IASs, and which have potentially devastating economic and ecological effects. It is often stated that plants with generalist pollination systems are more likely to become invasive because they can flexibly recruit a wide range of pollinators to fulfil their reproductive needs. But this clearly does not apply in the case of Tree Tobacco – in its native range, only hummingbirds pollinate the flowers. Insects may be excluded from the flowers because of the cocktail of chemicals that it contains. Research by Ido Izhaki and colleagues in Israel has demonstrated that the nectar of Tree Tobacco contains nicotine (not surprisingly) and the chemically similar anabasine. Both of these are powerful insecticides. Palestine Sunbirds, which pollinate the plant in the eastern Mediterranean, have a limited capacity to cope with these toxins, and the researchers suggest that, in the absence of alternative nectar sources, the birds may impose strong natural selection by remembering the location of plants that produce lower amounts of these alkaloids and choosing to visit these flowers preferentially.

However, the actual level of alkaloids in the flowers on a single plant may be variable and unpredictable. This has been shown in the related Coyote Tobacco, which is partially hummingbird-pollinated. The variability may in fact be a plant strategy to keep the birds moving between flowers: when they encounter a flower with high-nicotine nectar, they quickly leave, increasing the chances that more pollen from other flowers will be received on the stigma. For plants, as for animals, having a more diverse range of fathers improves the chances that at least one of them has a set of well-adapted genes.

Intriguingly, what capacity sunbirds have to deal with nicotine and anabasine seems to be mediated by the bacteria in their guts. The same is almost certainly true of hummingbirds – and once again we see the importance of microbes in the ecology and evolution of bird–flower relationships, as discussed in Chapter 10.

There are other examples of bird-pollinated plants that have become invasive, though they are not as common as invasive bee-pollinated or generalist plants. This reflects the relative scarcity of specialised bird pollination, however, rather than it being any particular barrier to a plant becoming invasive. Examples include a couple of species of fuchsia that use hummingbirds as pollen vectors in South America, one of which we encountered in Chapter 5. Then there are two large trees that share two distinctions: they have similar names, and their pollination ecology has never been studied in their native ranges.

Trees on fire

The epithet 'Flame' is applied to several tree species that have the trait of producing vivid red flowers which dominate the crown such that, from a distance, the tree appears to be on fire. From Madagascar comes Flame of the Forest (*Delonix regia*), a stunningly beautiful member of the pea and bean family that is widely planted across the tropics. Dry areas of eastern Africa are home to the African Tulip Tree (*Spathodea campanulata*), the original 'Flame Tree of Thika' that provided the title for Elspeth Huxley's 1959 memoir. As I said at the start of this book, I have kept the use of jaw-cracking scientific names to a minimum. Here, however, we start to see the value of standardised taxonomic epithets: two people discussing 'Flame Trees' could be talking about either of these species, or about members of the genera *Brachychiton*, *Nuytsia*, *Erythrina* or *Butea*, amongst others.

As well as sharing flamboyant flowers, many of these trees are, or are alleged to be, bird-pollinated. I've qualified that statement because while a few have been well studied, most have not. Despite their almost ubiquitous presence as ornamental plants across the warmer areas of the world, the pollination ecology of neither *Delonix* nor *Spathodea* has been researched in the areas where they evolved. There are published studies of the pollinators of both of these trees in India, where they are not native, and anecdotal observations of hummingbirds taking their nectar in the Americas. According to Gwilym Lewis's account of *Delonix* in *Curtis's Botanical Magazine*, the flowers are visited by Souimanga Sunbirds, which are common on Madagascar. But that's it. No one, as far as I'm aware, has conducted the detailed research in Madagascar and Africa, from where these trees originate, that would show which flower visitors are effective pollinators.

This is especially remarkable for *Spathodea*, which has been named as one of the 'World's 100 Worst' invasive species of any kind, plant or animal. *Delonix* is less of a global concern, though it is causing problems in Australia. Both of these Flame Trees form dense stands that dominate an area and exclude native species. Not only that, but there's experimental and anecdotal evidence that the pollen and nectar of *Spathodea* is toxic to bees, which may be a strategy to deter them from robbing the flowers, similar to that proposed for the Tree Tobacco – and the Canary Island Foxglove that I discussed in Chapter 6.

Why is it important that we understand the native ecologies of invasive species such as this? Why do we need to know their natural pollinators? One reason is that it can help us to predict which species are likely to become invasive in the future. As I emphasised earlier in the book, all organisms – be they plants, animals, fungi, bacteria or whatever – interact with other species throughout their lives, in relationships that include predation, parasitism, commensalism and the many and varied forms of mutualism. But when species are transported to a different part of the world, as has happened often during the Anthropocene, these interactions typically break down because usually only one of the participants moves. This loss of ecological relationships can play a role in whether a species becomes established in its new home.

Ecologists have mostly explored this issue from the perspective of the 'enemy release hypothesis' or ERH. The ERH predicts that, by leaving behind the predators or parasites or herbivores that normally limit its ability to reproduce, a species becomes more ecologically successful and ultimately invasive in its new range.

Potentially just as important is the 'missed mutualist hypothesis' (MMH), which in a sense is the mirror image of the ERH. As well as leaving behind 'enemies', introduced species leave behind 'friends' such as pollinators, seed dispersers, mycorrhizal fungi, defensive partners including some ants, and other mutually beneficial associates. Negative effects on reproduction arising from the loss of these relationships could potentially balance the positive impacts arising from the ERH.

The MMH is much less well studied than the ERH, which is why, during my time at the University of New South Wales as a visiting research fellow in 2019/20, I worked with Australian colleagues to review what was known about the MMH and suggest some fruitful lines of enquiry. The subsequent publication, led by Angela Moles, appeared in late 2022. It had a long gestation, with several iterations and revisions since we first started writing it, not least caused by the COVID-19 pandemic which hit just after I returned to the UK. But I think that the paper is all the better for it: sometimes slow(er) science is the best science.

Hopefully I've convinced you that, when it comes to invasive alien species, we need to look at the wider ecological context of how they are interacting with other species, especially mutualists such as pollinators. In this regard, as we saw in Chapter 9, a network approach can provide insights into their wider ecological roles.

Alien plant integration

Over the past 20 years or so I have been involved in some very productive collaborations with researchers in Brazil and Denmark, focusing particularly on hummingbirds and networks of plant–pollinator interactions. One of the outputs from this work, led by Pietro Maruyama, was published in 2016 in the journal *Diversity and Distributions*, and deals with the way in which non-native plant species are exploited by assemblages of hummingbirds. Our analysis used 21 plant–hummingbird networks, spanning the Americas, that had been studied by the authors or gleaned from the published literature. All of the birds were native (something that I will return to below) and the plants were categorised as either native or alien to the region where the study had been conducted. Using the kind of network statistics described in Chapter 9, we assessed the role of these non-native plants: do they behave the same or differently to the native plants? Do the birds distinguish between the native and non-native flowers? Does it matter whether or not they are adapted to bird pollination in their native regions?

In addition, for those networks where the data were available, we looked at how the number of flowers produced by a plant in the community, as well as hummingbird traits such as bill length, affected the way in which the birds interacted with alien plants. The results of the analyses were fascinating. There were 32 alien and 352 native plant species in the 21 networks as a whole, a ratio of 11:1 native to non-native.[1] However, the alien plants, on average, were visited by a greater number of hummingbird species and experienced a greater share of the flower visits, compared to native plants. Not only that, but the alien plants tended to be visited more often by some hummingbirds that were not observed visiting native plants. Surprisingly, none of the traits that were evaluated, nor floral abundance, were important determinants of the roles of these plants in networks. As far as the hummingbirds were concerned, it was short-billed species which more often visited the alien plants, rather than the long-billed birds.

Our main conclusion from the study was that once an alien plant is brought into the network of interactions, it becomes strongly integrated and, in some senses, may dominate the dynamics of the network. However, it's not generally possible to predict this integration from plant traits, and this may have wider consequences for understanding which types of alien plants become invasive, if these hummingbirds are their effective pollinators. Importantly, it's the short-billed hummingbirds, which tend to be

---

1 This ratio is similar to that found in other studies, for example the re-analysis of Charles Robertson's early-twentieth-century data by Jane Memmott and Nick Waser. Alien plants were 12.3% of species, all of which were visited to varying degrees by the native pollinators, including the Ruby-throated Hummingbird.

more functionally generalised as flower visitors and pollinators, that play the largest role when it comes to integrating alien plants into the networks.

Should any of this surprise us? Should we be amazed that hummingbirds will exploit novel nectar sources from plants that evolved in far-distant regions? On the one hand, no, it's just what we'd expect: hummingbirds are smart and, like all animals, they will take opportunities to gain food as and when it becomes available. Birds are especially adept at this, and can be quite flexible in their diets and creative in how they find food: for example, the Australian White Ibis is sometimes contemptuously referred to as a 'Bin Chicken' owing to its fondness for digging through trash in search of titbits.

On the other hand, these sorts of findings challenge the notion of hummingbirds as extreme specialists caught up in highly co-evolved relationships with plants that are finely adapted to pollination by their feathered partners. Although such textbook relationships do exist, they may not be representative of the broader ecology of hummingbirds and their flowers: by focusing just on the specialists, we may be missing some important aspects of the wider roles of these ecologically important birds.

But what of the flowers that find themselves displaced to different regions? How are they coping with this translocation? Quite well, it turns out, perhaps even to the extent that they are transforming into new species.

Foxgloves far from home

The major phases of European expansion and colonisation resulted in a two-way movement of species that has had ecological consequences ever since. Earlier in this chapter we saw that Tree Tobacco was moved, deliberately, from its native home in South America and now has become an important invasive alien species. That was a consequence of the European desire for novel plants with which to furnish their gardens. But for the colonists that made their homes in the New World, too much novelty was sometimes a bad thing and they longed for reminders of home. That was how the

Common Foxglove arrived in the Americas, far from where it had evolved in Europe.

When a team from the University of Sussex, led by Chris Mackin and Maria Clara 'Cala' Castellanos, studied the flowers of naturalised foxgloves in Colombia and Costa Rica, and compared them to native populations in southern England, they noticed something strange. The flowers of the New World plants were subtly, but significantly, different to those in the English populations. Specifically, the part of the tubular foxglove flower that becomes constricted and which serves to limit access to the nectar, was between 13% and 26% longer than in native plants. This doesn't sound like much, but such differences have evolved over a relatively short space of time – perhaps as little as 170 years – and they correspond to differences in the pollinators. If this continues into the future then the whole shape of the foxglove flower will change into one with a long, narrow end and a wider, flared opening, typical of many hummingbird flowers.

In England and the rest of Europe, the pollinators of Common Foxgloves are only bumblebees with quite long tongues. But in the Americas, foxgloves are pollinated by both bumblebees and hummingbirds. Crucially, Chris and Cala showed that the hummingbirds are even more effective than the bumblebees at depositing pollen, and it is they that are driving the evolution of the flower tube. The beaks of the birds are longer than the mouthparts of the bees and can reach further into the flower to access the nectar. In doing so, the birds leave more pollen on the stigma. As proof, the researchers showed that in most of the New World populations they examined they could detect natural selection in action. Common Foxgloves are mainly biennial, which is to say that they grow in the first year, flower in the second, then die. This rapid turnover of plants made it possible to show that those individuals with longer corolla tubes had greater seed production over their whole lifetimes.

There's still more work to be done on New World foxgloves and lots of unanswered questions, but the evidence is pointing to the flowers evolving in a rather different direction to the original population, under the influence of hummingbirds. It will be interesting to see if other flower traits, such as colour and nectar characteristics, also adapt to these new pollinators.

At this juncture you might be wondering what has happened to the 'bad birds' that the chapter title promised. Well, the truth is that there aren't that many of them to discuss, certainly not compared to invasive plants. Why that might be the case is an interesting question.

Bring on the birds

Birds have been kept in cages and aviaries for thousands of years, and moved around the world for food or sport or simply (as with the foxgloves) to remind colonists of home. Consequently, there's no shortage of species that have been transported far from their natural homes and which flourish in the regions where they find themselves. Ubiquitous examples are the House Sparrow and the Common Starling.

As far as I know, none of these translocations have involved either hummingbirds or sunbirds, perhaps because their nectar-dominated diet makes them difficult to keep in captivity outside of zoos. The only specialist flower-feeders that have established feral populations are honeyeaters.[2] Even here, however, their status as invasives is rather different to that of the plants we have been considering. Reviewing the topic of the 'invasion ecology of honeyeaters', Janette Norman and Leslie Christidis were mainly concerned with how these birds have spread of their own volition from Australia and New Guinea eastwards into the islands of the Pacific. Some of this has happened in historical times. For example Long Island, about 60 kilometres from the coast of New Guinea, suffered a catastrophic volcanic eruption around 300 years ago which wiped out all of the plants and animals. Since then, around 50 species of birds have recolonised, including the Bismarck Black Myzomela and Sclater's Myzomela. Jared Diamond refers to these as 'supertramp honeyeaters' because they are able to spread back into islands that have been devastated by volcanoes or tidal waves.

2 If others exist, I'm sure someone will tell me!

The key seems to be the high population density that such birds can achieve very rapidly. Other honeyeaters can do the same, and at least one is considered to be a 'bad bird' in the sense of this chapter title.

The Noisy Miner is a honeyeater that lives up to its name – we shared a garden with a pair in Australia and they make quite a racket, as well as being notoriously aggressive (Plate 28). Noisy Miners are native to southeastern Australia, and under normal circumstances they survive in forest edge and disturbed habitats. The circumstances in that part of the continent are not normal, however. Forest clearance over the past couple of hundred years has left the formerly contiguous old-growth forest reduced to large numbers of much smaller patches, just the kind of habitat that the Noisy Miner prefers. They have now become 'overabundant' and subject to lethal control by wildlife specialists, because at high densities these honeyeaters, which live in cooperative social groups, can aggressively displace other birds.

In most areas where Noisy Miners have been culled, the abundance and diversity of other species has increased. Although it's clearly effective, how feasible is culling as a long-term strategy? Would it not be better to stop deforestation, restore areas that have been cleared, and join together smaller fragments in order to reduce the preferred habitat of Noisy Miners? These are only 'bad birds' because human actions have disturbed the normal checks and balances on the sizes of their populations. In fact, recent work by Jade Fountain and Paul McDonald at the University of New England, Armidale, concluded that they are more aggressive in gardens where people have planted abundant nectar sources. In an accompanying commentary in *The Conversation*, the authors provide guidance on how appropriate gardening practices can reduce the impact of Noisy Miners on other birds.

In Australia, and elsewhere, there are plenty of examples of invasive generalist birds that are occasional flower feeders and pollinators. We encountered two in Chapter 15: Eurasian Blackbird and Common Myna pollinating feijoa in New Zealand. Various introduced myna species are common across Asia and in Australia too, and often visit flowers. Likewise, doves and parakeets are becoming more common outside of their home regions, in part because they are such popular aviary birds. In Britain, the Rose-ringed (or Ring-necked) Parakeet has become a fairly common sight in the parks and gardens of our larger towns and cities. There is an established colony in a park close to where we lived in Northampton and I'd occasionally see them dashing through our garden at high speed. They feed mainly on fruits and seeds, but may also consume flowers and nectar. Their role as a pollinator in their native range of south Asia and central Africa has never been established, though it's not impossible, as other members of the Old World parrot family are certainly known to effectively transport pollen.

The Rose-ringed Parakeet turns up time and again in reports of non-native birds in other regions, for example in the review of the invasive birds of Hong Kong by Michael Leven and Richard Corlett. Those researchers also noted that two specialist flower-feeders – Fire-breasted Flowerpecker and Fork-tailed Sunbird – seem to have colonised from southern China at some point in the 1950s. Their consequent spread across the region is attributed to habitat change, including maturation of restored forests and possibly planting of garden ornamentals. As with the colonising honeyeaters mentioned above, whether we should call these species 'invasive' is debatable, as their appearance in Hong Kong seems to be partly by indirect human agency and partly by natural processes. However they are certainly representative of the wider issue of how humans affect birds and their interactions with flowers, for better or for worse.

In contrast, the Warbling White-eye now finds itself very far from home, on all of the islands that make up the Hawaiian archipelago. There it visits the flowers of a wide range of species for nectar and has become a conservation concern because it seems to be out-competing native birds, including one of the endemic honeycreepers, the Hawaii Akepa. Invasive alien species, including plants and animals, are just one of many threats to the relationship between birds and flowers. In the next chapter I'll explore some of these threats in more detail.

CHAPTER 18

What escapes the eye

The image in Plate 4, showing a Rainbow Lorikeet visiting the inflorescence of a banksia tree in a suburban Sydney street, has a strange, yellowish hue, almost as if it's been washed in nicotine. That's not due to any fault of my camera phone, it reflects the quality of the light that we experienced in that part of Australia in late 2019 and early 2020. Throughout the southeastern part of the continent, bushfires were raging uncontrollably, daylight was filtering through smoke, and fine ash fell from the skies. Parts of Australia have always annually burned to some extent, but that fire season was different: the burns were bigger, hotter, more extensive than normal. Exacerbated by drought and strong winds, the bushfires were, if not a product of climate change, then strongly influenced by them.

If, as seems entirely plausible, birds have been pollinating flowers for at least 50 million years, a big chunk of that evolutionary history has occurred in the ancient land mass that we now think of as Australia. There's fossil evidence of banksias from at least 60 million years ago, and while we can't be sure that they were bird-pollinated, we do know that the Old World parrots (to which lorikeets belong) are also an ancient group. Similarly, the honeyeaters are often described as having 'deep phylogenetic roots' within the passerines, and as we saw in Chapter 1, they have a fossil record extending back to at least the Miocene.

Human domination of the planet through habitat loss, pollution, introduction of invasive species and climate change has resulted in what's been termed a 'biodiversity crisis' or the 'sixth mass extinction'. Extinction is not just the loss of a species, however, it's the loss of evolutionary history and the loss of the interactions in which that species was engaged.

Such losses have happened, naturally, many times over the past 50 million years, because species do not live forever. Of course, there are examples of 'living fossils', species that have persisted, largely unchanged, for tens or even hundreds of millions of years. The fact is, however, that on average most vertebrate species exist for only a few million years, after which they go extinct for a variety of reasons. These could be abrupt changes in the climate or some other aspect of the environment, the evolution of a particularly virulent microbe, or geological upheaval of terrestrial or extraterrestrial origin, such as volcanoes or asteroids. Whatever the reason, species have a finite lifespan. It is estimated that 90% of all the species that have ever existed are now extinct.

This 'background' level of species extinction applies to the human lineage too; of the perhaps 10 or so species or subspecies of the genus *Homo* that have existed over the past 2 million years, only one, *Homo sapiens*, still lives. And how we live! In extreme abundance – 8 billion people and rising – and with a big impact on the planet. Depending on how you calculate the figures, perhaps 20% of the land surface is dominated by farming and urbanisation, 30% of marine food stocks are overexploited, and more than 20% of the world's net primary productivity is directly or indirectly exploited by people.

Human impacts on the environment are varied, but perhaps most compelling in their severity when we consider extinction rates. Today, the rate of species loss is estimated to be between 1,000 and 10,000 times higher than the background level. However, 'extinction' can be both global (the species has gone forever) or regional (it has been lost from a country or part of a country), which we term 'extirpation'. Factors such as overexploitation, loss of habitat, pollution and biocides, and more recently anthropogenic climate change, have all conspired to reduce the level of biodiversity, globally and regionally, below its historical levels. This is true across all branches of the tree of life, and birds are no exception.

In September 2022 the conservation organisation BirdLife International provided an update of their *State of the World's Birds* report that made for profoundly dismal reading. One hundred and fifty-nine species are known to have become extinct due to human activity, with another five extinct in the wild but still living in captivity. In addition, a total of 672 species were classified as Endangered or Critically Endangered, and 773 are Vulnerable. These are species that have had population declines of some kind, often with dramatically reduced distributions compared to what they were historically. Of the 11,000 or so living or recently extinct bird species, therefore, about 15% have been severely impacted by humans.

Extinction or extirpation of a species grabs the headlines, but we often forget that the loss of one species can have profound consequences for the other organisms with which it is connected in a complex set of relationships. Interactions that have evolved over millions of years, and in some cases resulted in co-dependency between two or more species, are disrupted. What happens when one of the partners is lost from a community? What if a bird loses an important nectar source or, more likely, a plant loses its bird pollinator? When a bird goes extinct, so too do the specialised bacteria, fungi and parasites with which it's associated. And so too might the plants that depended on it for pollination and seed dispersal. As ecologist Dan Janzen put it, back in 1974:

> What escapes the eye, however, is a much more insidious kind of extinction: the extinction of ecological interactions.

And it's not just global extinction that Dan was referring to; he had in mind extirpations too. Because if we lose a pollinating bird from an ecosystem, we lose its functional role. Before we consider this, however, let's look at how pollinating birds are faring in this *Homo*-dominated world.

Extinction is forever, but extirpation is pretty bad too

Make no mistake, the idea of regenerating extinct species is pure fantasy in the realms of *Jurassic Park*.[1] Extinct species are gone,

---

1 Current efforts to resurrect extinct species are fraught with technical issues and raise huge ethical concerns – see, for example, the Wikipedia article on Woolly Mammoth.

whether the cause was humanity or natural processes. According to the wonderful *Birds of the World* website, approximately 15% of the described hummingbirds (that's 50 species) 'face serious conservation challenges'. Since the mid-nineteenth century, two species are known to have become extinct. A third species, the Turquoise-throated Puffleg from Ecuador, was last observed in the 1970s, though that record is unconfirmed and ornithologists fear that it may also be extinct. The two hummingbirds which are definitely gone were both Caribbean species. As we saw in Chapter 12, island species, of all kinds, are often the most threatened. Elsewhere there's concern about species such as the Juan Fernandez Firecrown, which is found only on one of the tiny islands in the Pacific archipelago of that name. Like so many insular species, its population is declining because of the usual culprits of ongoing degradation of its habitat and the depredations of invasive mammals. These pressures, especially destruction or alteration of natural habitats, combined with limited natural ranges and tiny population sizes, are also affecting those other 15% of endangered hummingbirds.

The conservation statistics for the other major groups of bird pollinators vary in their levels of concern. In order to understand how species are faring in our ever-changing world, the International Union for Conservation of Nature (IUCN) developed a set of criteria to classify species as Vulnerable, Endangered, Least Concern, and so forth. According to those criteria, in the sunbirds and spiderhunters, about 5% are Vulnerable or Endangered, whilst none are Critically Endangered or are known to have gone extinct. Some 5% are Near Threatened, but the remaining 80% are Least Concern, and we can say that with some certainty because this family has been so well studied that none are Data Deficient.

It appears that Africa and Asia are holding on to their main family of pollinating birds a little better than the Americas. That's no cause for complacency, though, because three species are classified as Endangered, just hanging on by their claws: the Elegant Sunbird, found only on the small island of Sangihe (or Sangir) north of Sulawesi, an area prone to volcanic disturbances, and the East African endemic Loveridge's and Amani Sunbirds. The latter two species live in rather small pockets of montane forest. I've had the privilege of seeing the Amani Sunbird in the area from which it derives its name, the Amani Nature Reserve in the East Usambara Mountains of northern Tanzania, taking nectar from flowers of both native and non-native plants. The latter are extremely abundant. owing to the proximity of a colonial-era botanic garden that introduced tropical rainforest plants from across the world, because apparently the almost 3,500 species of native plants in the Usambaras were not enough to maintain the interest of botanists.

The honeyeaters are likewise mainly doing okay across Australasia and the western Pacific, with 84% being Least Concern. However, 14 species (8% of the total) are at conservation risk, with one being Critically Endangered: introduced rats are taking their toll of New Caledonia's Crow Honeyeater. One species, the Chatham Island Bellbird, was last seen alive in 1906 and is almost certainly extinct: once again, it's island birds that bear the brunt of human impacts.

Loss of old-growth forest is having an important effect on the abundance of some honeyeater species in Australia itself, as we saw in the previous chapter, exacerbated recently by the more extreme bushfires that the southeast, in particular, is being exposed to. There's also a longstanding concern that an over-abundance of introduced, feral Western Honey Bees could be negatively impacting the native birds that rely on the same nectar and pollen. Of particular concern is the Regent Honeyeater, a once widespread species found from Adelaide to southern Queensland, now reduced to smaller, scattered populations. It was Critically Endangered even before the 2019/20 fires season, which affected a number of the remnant strongholds of the species, with a 2018 study ranking it the seventh most likely Australian bird to go extinct. One of the consequences of having such small local populations has been a decline in male song complexity, which is impacting reproductive success: males which sing the wrong song for a local area, or even use the songs of other species, are less likely to find a mate.

In contrast to the sunbirds and honeyeaters, the white-eyes and their relatives are, as a group, a greater conservation worry, even more so than the hummingbirds. Only 54% of the 142 species are of

Least Concern and one-third are threatened to some extent, with five species Critically Endangered. One of these is the Sangihe White-eye, endemic to the island that's also home to the endangered Elegant Sunbird. The original type specimen of this species was collected in 1886, but it was not seen again by ornithologists until 1996. It's estimated that its population may consist of fewer than 50 birds.

Another Critically Endangered species is the White-chested White-eye, endemic to Norfolk Island and now thought to number no more than 20 individuals over an area of 5 square kilometres. There's a strong probability that it may even be extinct. As so often, introduced species (mainly rats and cats) are responsible for its demise, though competition with the Silvereye has also been implicated. The latter species, which is a member of the same family, underwent an irruptive expansion from Australia to New Zealand in the nineteenth century, colonising the isolated Norfolk Island around 1900.

Unusually amongst the major groups of flower-visiting birds, over-collection of Asian white-eyes to sell as caged songbirds is also having a huge effect on some wild populations.

It would be possible to go through all of the families that are included in the table in Chapter 3 and provide similar statistics, but that would make for rather dull and often dismal reading. I do want to highlight one final group of pollinating birds, however, about which it's too late to be concerned: the Hawaiian honeyeaters (Mohoidae), all five species of which are extinct. Despite their name, coined before the advent of molecular phylogenetics, these birds are (were) members of the clade that includes the waxwings and are not closely related to the other honeyeaters. They form(ed) a distinct family in their own right, and it's the only example we have of an entire family going extinct due to human pressures on the environment. Three species became extinct in the nineteenth century, one may have held on until the early twentieth century, but the final species, the Kauai Oo, was last recorded in 1987. These birds were under pressure from the indigenous Hawaiian people even before the arrival of Europeans because they were extensively hunted to provide feathers for the head-dresses, cloaks and other accoutrements worn by the nobility. Feather collecting had likely been going on for hundreds of years since the first colonisation of the islands in around 1000 CE, but it's debatable whether this was sustainable, as we know of at least one and possibly two species of Hawaiian honeyeaters that had been driven to extinction prior to the arrival of Captain James Cook in 1778. The Hawaiians had brought with them, intentionally or otherwise, mammals such as rats and pigs, to which the Europeans added more, intensified the destruction of habitats, introduced alien birds and mosquitoes (and avian malaria), and ultimately doomed these island endemic birds.

As if that weren't bad enough, another set of flower-visiting birds, the Hawaiian honeycreepers (members of the finch family) have also been extensively exterminated from the islands: at least 20 species have become extinct since European contact, with a similar number of known extinctions dating from the first arrival of people. The remaining species, which number around 17 or so, are almost all endangered to some degree, and several may actually be extinct. Not all of the species in this group are (or were) nectarivorous, but at least 15 were and many of these possess long, thin, downcurved beaks to allow them access to nectar at the bottom of deep flowers.

Which brings us back to Dan Janzen's words. We know from contemporary accounts that Hawaiian honeycreepers included a lot of nectar in their diets, and nineteenth-century paintings as well as the few twentieth-century photographs often depict them close to flowers. Were they important as pollinators? The answer is almost certainly yes, and the best evidence is provided by the spectacular diversity of a clade of plants that, like the birds, is endemic to the Hawaiian islands.

When the birds are gone

The Hawaiian lobelioids (or lobeliads) are members of the bellflower family (Campanulaceae) and, as the name suggests, they are related to the familiar garden lobelias. They have undergone a spectacular evolutionary diversification – what biologists refer to as a 'radiation' – into somewhere between 125 and 140 species (taxonomists differ in their conclusions). Many of these plants are bird-pollinated. Indeed, there is a good match between the curved flowers of some lobelioids and the curved bills of some birds, which suggests a relationship similar to that of the hummingbird–heliconia story I described in Chapter 11. Not only that, but nectar traits including low sugar concentration and dominance by hexoses in the 22 species of *Clermontia* studied by Richard Pender and colleagues strongly implies that these evolved to be bird-pollinated. There are also species within the radiation that are autogamous (i.e. can produce seeds in the absence of pollinators), whilst others have co-opted non-native white-eyes as effective pollinators. That at least gives us hope that the reproduction of some of these plants may continue into the future, even in the absence of the pollinators with which they evolved.

Other endemic plants, however, may not be so fortunate. The endemic Hawaiian mint studied by Clare Aslan is also likely to have been originally pollinated by birds. Although the plant can produce some seeds by self-pollination, and non-native bees make extremely rare visits to flowers, hand-pollinating the flowers to mimic what the birds would have achieved increases average seed set at least three-fold. The reproductive assurance gained by the seeds that result from self-pollination means that this species is not likely to go extinct, though it limits the genetic variation of an individual's offspring. As far as the wider network of flowers and pollinators is concerned, though, the species is ecologically extinct and, to quote the authors, 'today contributing little to pollination networks in its high-elevation habitat'. In this sense we can think of the plant as being 'functionally extinct', an issue that goes beyond concerns about extinction or extirpation and which may be more common within ecosystems under pressure than we currently realise.

Intriguingly, research in the 1990s showed that the beak of one honeycreeper, the Iiwi, has become slightly shorter and less curved as a result of natural selection favouring birds that can feed on flowers with straighter corollas than those possessed by the lobelioids. This might be because the plants have such low population sizes that they are being overlooked even by birds that could pollinate them. There are other examples, including on the Canary

Islands where the bird-pollinated endemic bird's-foot-trefoils have extremely small populations and are largely ignored by their pollinators except in some parks and botanic gardens. For most bird-pollinated plants across the world we don't have this kind of information. Although the IUCN criteria are applied to plants, the species accounts are often unaccompanied by any consideration of what their pollinators might be, because most plants have not been studied in detail.

Dealing with uncertainty

One of the advantages of the IUCN's system is that it gives us a useful shorthand with which to categorise species with respect to their population trends in relation to existential threats.[2] For those species that have been assessed, this gives us some degree of certainty about which require special conservation efforts and which are doing just fine. For flower-visiting birds, we have some quite detailed data on population trends and distributions for particular, often charismatic, species (for example hummingbirds) or for certain geographic localities or regions. But for most species of pollinators, the insects that dominate flower–pollination networks, we know almost nothing. In my book *Pollinators & Pollination* I discussed the current state of our knowledge of how populations of pollinators have changed over time. As I wrote in the chapter 'The shifting fates of pollinators', most species are Data Deficient, which is to say that we don't know how, or whether, their populations are changing. It may be that they are doing well, but they could also be declining. This is particularly true for those regions of the world that hold the greatest terrestrial biodiversity: the tropics. For the vast majority of species in the tropics we know precious little about trends in their populations and how their distributions have changed in the face of wide-scale land transformation and recent climatic shifts.

---

2 The IUCN categories do not necessarily reflect a species' rarity, as is often presumed. Many species are naturally rare and have low-density populations, sometimes scattered over a small area, but are not threatened with extinction.

Filling in some of the gaps in our knowledge of Neotropical pollinator distributions was one of the aims of a project in which I was recently involved called SURPASS2. This was a collaboration between South American and UK ecologists who were interested in pollinators and their conservation. In one of the research papers that came out of that work, led by Rob Boyd from the UK Centre for Ecology and Hydrology, we used the *Global Biodiversity Information Facility* (usually shortened to GBIF) database to look at the changing distributions of some pollinators. Since its inception in 2001 as a central repository for records of where species occur, GBIF has grown into an important resource for scientists who are interested in mapping and conserving biodiversity. However, it is not complete and it is not perfect. We were particularly interested in understanding the kinds of biases that come with such publicly available data, and whether recent efforts to add data to GBIF have improved our understanding of trends in pollinator populations.

Our study focused on four important groups of Neotropical pollinators: bees, hoverflies, leaf-nosed bats and hummingbirds. Not surprisingly, hummingbirds were well represented within the data set, for several reasons. They are not especially diverse (although they are the second or third largest family of birds),[3] and their relatively large size makes them easier to record in comparison to the insects. As interest in these birds, and access to their habitats, grew from the 1950s onwards, so there has been a decade-on-decade increase in the number of GBIF records for hummingbirds, such that in the most recent time period (2000–2019) there are observations for almost all hummingbird species found in the Neotropics, compared to about 70% in the 1950s. This in itself presents problems for understanding population trends: how were those 30% faring 70 years ago?

Another bias within the data set (for all species, not just hummingbirds) is that there tends to be an over-representation of rare species, especially in more recent years. This reflects the

---

3 In comparison to the *c.*360 species of hummingbirds, we estimate that there are at least 2,000 Neotropical hoverflies and 5,000 bees.

behaviour of naturalists – who are keen to go and see the least common and least well-known species, perhaps at the expense of submitting records of the easy-to-see and widespread species. Finally, our assessment showed that the available information for the hummingbirds is significantly geographically biased, with an over-representation of records from the Andes in Ecuador and Colombia, compared with the Amazon Basin, for instance. Again, this reflects the behaviour of birdwatchers who wish to see the mountain endemics rather than the more widespread lowland species. Our overall conclusion is that there are significant limitations and biases inherent in all of these data sets even for groups like hummingbirds which one would imagine are well documented by scientists and birdwatching naturalists. This is something that needs to be taken into consideration whenever we describe how well or how badly a species is coping with the environmental changes that humans are imposing on the natural world.

There's another reason why this kind of research is important. For as long as people have been watching birds, they have noticed that their distributions are not fixed: once common birds become rare, whilst previously unseen birds suddenly appear. Recently, the assumption has been that such changes are due to human activities, but that does not have to be the case. There are lots of examples of birds that have historically expanded their distributions across Europe, such as Eurasian Collared-Dove, Little Egret, and Cetti's Warbler. Flower-visiting birds are not exempt from these kinds of range shifts, as we saw in the previous chapter, and those examples are certainly not unique. In North America, the winter range of Anna's Hummingbird has moved more than 700 kilometres northwards, likely the result of increased urbanisation and the provision of sugar-water feeders by residents. There are other, non-human causes of changes in bird distributions, however.

The avifauna of central Victoria has been well documented by Australian naturalist Geoff Park on his *Natural Newstead* blog. I had the pleasure of accompanying him for a day's birding when it transpired that a close friend of mine also lived in Newstead, and I can attest to how knowledgeable and careful Geoff is as an observer of the natural world. In a blog post in September 2022 he noted:

> Once considered an absolute rarity in central Victoria, the Scarlet Honeyeater ... is now regularly observed ... In recent years it has been increasingly reported around Melbourne (where it was once a rare visitor) and further west.

As you can see in Plate 25, the Scarlet Honeyeater (also called the Scarlet Myzomela) is a stunningly coloured, diminutive bird. It's normally resident mainly further north in subtropical and tropical Australia, with southern populations migrating south as nectar sources bloom. It's also known to be an irruptive species, with sudden influxes appearing in areas where it's rarely if ever encountered. Irruptive behaviour may be the key to the spread of certain species that are able to rapidly colonise new areas, an argument I've made for the Tree Bumblebee, which only arrived in Britain in 2001 and is now one of the commonest bees in the country.[4] It remains to be seen whether the Scarlet Honeyeater will continue this expansion of its range.

The assumption for the Scarlet Honeyeater is that unpredictable and long-term drought, characteristic of the Australian environment, causes an absence of flowers, which forces these nomadic movements. More recently, species in other parts of the world are being subjected to the effects of climate change, mirroring what Australia has long experienced, with extreme weather driving mass movements.

Alternatively, more benign conditions can allow a species to enter landscapes from which it was previously excluded. Both of these processes are happening simultaneously, and we are currently going through a global experiment in which species are moving into novel regions, of their own accord or via the actions of people. One of the outcomes of this experiment is that the card deck of

---

4 See pages 182–188 of *Pollinators & Pollination*.

potentially interacting species gets shuffled, and new relationships may emerge, in our case between populations of birds and flowers that have not previously encountered one another. That's fine if it's a non-disruptive novel encounter, as seems to be the case for the Tree Bumblebee: as far as we can tell, no other bees have been displaced and it's an effective pollinator of some plants. However, it may be that the range shift or the increase in a bird's population leads to competition or aggression, as we saw for the Noisy Miner. That's a more problematic situation.

The consequences of a loss of bird pollinator diversity to a region can be profound, especially if a high proportion of plants relies on them for reproduction, as is the case in New Zealand. Long-term declining populations of bird-pollinated species are likely to be replaced by other plants that are not so bird-engaged. Ultimately, plant communities change – and with them, potentially, the nature of those ecosystems. That said, I don't want to finish this book on a pessimistic note and leave the reader thinking that all is lost, and that there's no hope for pollinating birds, the plants which depend on them, and the ecosystems in which they are embedded. Conservationists, scientists, governments and individuals are working extremely hard to counter that past centuries of environmental degradation, and it is these glimmers of hope that we will discuss in the final chapter.

CHAPTER 19

The restoration of hope

Optimism is infectious. During the decades I have spent as a researcher and educator, it has been the optimism of the scientists trying to understand the natural world and the practitioners aiming to conserve it that has kept me buoyant. Even at my lowest ebb, when all the evidence points towards humanity's destructive and self-defeating impulses, it's the optimism of such people that sustains me. There are some environmentalists who see this as naive, head-in-the-sand denial of the facts and insist that angry pessimism and a sense of despair are the only appropriate responses. They may be correct. Society may be on the verge of environmental collapse. As I write, we're seeing reports that the Earth's oceans are experiencing unprecedented warming which could have unpredictable and catastrophic consequences. Life-threatening heatwaves and associated wildfires are occurring across some parts of the western hemisphere, devastating flooding is occurring in Asia, and the predictions are that these will get worse in the coming years. But that doesn't mean we should adopt a position of 'doomism', as climate scientist Michael Mann calls it. As the originator of the famous 'hockey stick' graph, Mike has spent the last 30 years trying to convince the public and politicians that anthropogenic global warming is both real and potentially more devastating than we realise. This has drawn criticism, insults and attempts to discredit him from large corporations, state actors and online trolls. He probably has more reason to be pessimistic than most, and yet I discovered that his optimism is one of his defining characteristics, even though he knows better than most how high the stakes are.

In an odd but somehow concordant coincidence, Mike was a visiting researcher at the University of New South Wales at the same time as me and, like me, was writing his next book, which appeared in 2021 as *The New Climate War*. We'd not previously met, and in fact I only discovered that he was in the area when he tweeted a photograph of a view that was by then familiar: ash falling through the smoky air of Coogee Bay. After I introduced myself we met up a few times to discuss science and politics, our chats soundtracked by the persistent calling of Noisy Miners, and have stayed in touch since. These meetings convinced me that pessimism is not a healthy attitude, and that if we give into it, 'they' have won. If Mike can retain his optimism in the face of both environmental and personal abuse, then we should quit with the doomism and turn our positive faces to the world of conservation.

The previous chapter, about the impacts humans have had on birds and flowers, makes for sad reading. But there are conservation success stories out there too. Globally, millions of citizens are mobilised to do what they can for species and for ecosystems. Flower visitors such as hummingbirds, honeyeaters and sunbirds, with their vibrant colours and often confiding attitudes to the presence of people, can certainly act as flagship species for these conservation efforts, and there are plenty of examples to discuss. In this chapter I want to highlight some of the success stories when it comes to conserving birds as pollinators and their interactions with flowers.

The period 2021 to 2030 is the United Nations' Decade on Ecosystem Restoration, described as

> a global rallying cry to heal our planet … [that] … aims to prevent, halt and reverse the degradation of ecosystems on every continent and in every ocean. It can help to end poverty, combat climate change and prevent a mass extinction.

Restoration of ecosystems is about much more than planting trees or sowing wildflower seeds in grasslands, however. Ecosystems are about the processes that make the natural world function, including nutrient cycles, water capture, carbon sequestration and soil formation. Living organisms are required for all of these, of course, but more critically it's the interactions *between* the species that make the ecosystems work. Pollination by animals such as birds, of course, is an important example of such an interaction, and that's one of the reason why their conservation is so important.

Conservationists all over the world are focused on preserving these ecologically important birds. In South Africa, Anton Pauw has instigated the Ingcungcu Sunbird Restoration Project, named for the Xhosa word meaning 'long-billed bird'. The term also refers to royalty, which is fitting for such a regal avian group. The project aims to 'heal the relationship between plants, birds and people by restoring migration routes for nectar feeding birds across the city' of Cape Town. There we have that word – 'heal' – again, an important way of thinking about how people can reverse the harm humanity has caused and for which we all bear some responsibility. Ingcungcu has had some success in this regard by restoring habitats and engaging young people in the work. One of their initiatives, led by Bongani Mnisi, enabled the pupils of eight schools to plant indigenous, bird-pollinated species which served to link up two existing conservation areas. Over a seven-year period, the diversity and abundance of sunbirds increased as the plants matured, and the school gardens acted as stepping-stones between the conservation sites.

There are many smaller-scale initiatives too, similar to Jonas's garden in Brazil (see Chapter 16). On the Caribbean island of Trinidad, retired agricultural scientist Theo Ferguson and his wife Gloria have created an 'ashram for the hummingbird' by planting appropriate flowers and hanging sugar-water feeders in their garden. They've named their property *Yerette*, an indigenous word for 'home of the hummingbird', and an appropriate moniker for a garden that regularly attracts 15 species.

Notwithstanding some of the concerns about sugar-water feeders that were raised earlier, there are many such examples of homeowners who are supporting hummingbirds across the Americas. Such initiatives are locally important for the birds – as well as highlighting people's commitment to their conservation. However, if we are to reverse habitat loss, support efforts to combat climate change, and conserve birds and their flowers, we need large-scale projects involving networks of committed citizens. And this is entirely possible if there's the political will to do so and if we include local people in the plans from the beginning. Starting in the 1990s, Nepal allowed local community groups to take over the forests on which they and many flower-visiting birds depend (see Chapter 12). The result was an almost doubling of forest cover, from 26% to 45%, over a 25-year period. These community forests are the responsibility of more than 22,000 community groups, now cover 2.3 million hectares of Nepal, and account for about one-third of the country's afforested area. It's a marvellous achievement, and a testament to that country's government and its peoples.

Bringing back the Hawaiian honeycreepers

As we saw in the previous chapter, the two main groups of pollinating birds in Hawaii have suffered more than any other from human interference in their ecosystems: the endemic honeyeaters are all extinct, and the honeycreepers have been reduced to about half their original diversity. Although it's too late for the Hawaiian honeyeaters, the honeycreepers are the subject of vigorous conservation efforts that are proving successful.

The species that are still extant are predominantly found at higher elevations, for two reasons. First of all, it's mainly lower-elevation habitats that have been deforested and converted to agriculture, ranching and urban developments. Secondly, the mosquitoes that pass on avian malaria cannot thrive at the lower temperatures found higher in the mountains, though there is a concern that this may be altering as global climate change increases average temperatures. Removing invasive alien plants to improve the quality of the native habitats is an important part of the conservation strategy for these birds, though several studies have shown that honeycreepers take nectar from some of these plants. In 2022, for example, writing in the *Journal of Pollination Ecology*, Seana Walsh, Richard Pender and Noah Gomes proposed that two honeycreeper species, the Iiwi and the Hawaii Amakihi, were obtaining a significant proportion of their daily energy requirements by robbing nectar from the long flowers of an invasive passionflower.

On balance, though, it's undoubtedly better to have large expanses of high-quality habitat that can support these birds (and other wildlife) than to have them rely on outsider plants. And the Hawaiian islands are certainly pushing forward in their effort to restore highly degraded landscapes, as shown by the many individual projects that have their own websites. For example, the Maui Forest Bird Recovery Project (MFBRP), based in the 614-hectare Nakula Natural Area Reserve, is trialling various approaches to exclude non-native large mammals and replant the native vegetation. The MFBRP works with volunteers to collect seeds, remove invasives and monitor progress, a sound approach because having local people devote their time and energy to such conservation projects is often key to their success. This is true for all types of conservation efforts, all over the world, not just in Hawaii, and not just forests and birds.

But volunteers are not enough, and it's always gratifying to see governments (at all levels) or businesses get behind such initiatives. It was satisfying therefore that in May 2022 there was a Hawaii state press release announcing:

> RACE TO SAVE HAWAIIAN HONEYCREEPERS BOLSTERED BY $14 MILLION IN FEDERAL INFRASTRUCTURE AID.

This is a significant boost to the conservation efforts for these iconic birds – and I've left the original capitalisation because such good news deserves to be shouted about. Chris Farmer from the American Bird Conservancy was quoted as saying:

> We have a huge group of scientists, conservationists, land managers, and others from non-profit, private, and government organisations engaged in a collaborative process to break the avian disease cycle and save these birds as quickly as we can. We can only do that by

controlling non-native mosquitoes where our honeycreepers have their last mountain refuges.

This last point is crucial, because there's no point in restoring and protecting habitats (or releasing captive-bred birds, another string to the conservation bow) if those habitats are still home to invasive species that have a big negative effect on the population of a bird. Removal of the mosquito problem, and therefore the avian malaria, is being tackled on multiple fronts, including control using chemicals and releasing sterile male mosquitoes into the wild. It's also heartening to learn that at least one species – the Hawaii Amakihi – seems to have evolved a tolerance to the malaria and is expanding its population into the lowlands, where, as I noted, it is a generalist nectar feeder on both native and introduced plants.

The Hawaiian honeycreeper story highlights two aspects of ecology that make for a successful conservation project: it is necessary to remove the threat(s) to a species (in this case avian malaria and cats), and to ensure that the target organisms have sufficient habitat in which to thrive. It's also important to track the progress of such programmes so as to determine whether they have been successful.

The impact of habitat restoration on birds and bird pollination

In a recent review and meta-analysis of how habitat restoration impacts pollinators and pollination, Luísa Genes and Rodolfo Dirzo assessed 26 studies from Australia, Brazil, Colombia, Costa Rica, Mauritius, Seychelles, Sweden, UK and USA. Overall they concluded that when degraded sites were restored, the levels of pollination increased up to that of undisturbed sites. Not surprisingly, given what was said in Chapter 4 about the relatively low frequency of birds as pollinators, most of the pollination was done by insects. However, birds featured in some of the tropical studies, and a couple of them are worth considering in detail.

Working in restored forest sites in Costa Rica, Catherine Lindell and Ginger Thurston looked at how three major factors – the amount of flower resources in an area, the design of restoration planting, and the forest cover in the landscape – affected the number of pollinating bird visits to the Guaba tree. This is a species widely used in restoration schemes and planted as a crop for its edible fruit pulp, hence the nickname 'the ice-cream bean'. The researchers observed visits by nine species of avian pollinators: seven hummingbirds, the Tennessee Warbler and the Bananaquit. They discovered, perhaps not surprisingly, that those trees with more flowers had more pollinator visits. What is more surprising is that restored sites adjacent to larger areas of existing forest cover did not attract more birds. This suggests that the birds will find such sites, and pollinate the trees, even in more degraded landscapes. Forest restoration is worthwhile regardless of the ecological history of a region.

Elsewhere in the world we can find similar stories of researchers assessing the reproductive success of bird-pollinated plants in restored sites and showing that conservation interventions work. When they studied a species of banksia on three post-mining sites in Western Australia, ecologists Karen Frick, Alison Ritchie and Siegfried Krauss found similar levels of genetic outcrossing in the seeds and the same overall pollination services, compared to natural stands of these trees. They concluded that the patches of restored banksia 'are resilient to human impacts, due largely to their generalist pollinator requirements and highly mobile avian pollinators'.

The same kind of results are found when tropical habitats are restored using a different approach, the removal of invasive alien species, as ecologist Christopher Kaiser-Bunbury and colleagues found when working in highly degraded and restored habitats in Mauritius and the Seychelles. Habitat restoration resulted in a significant increase in the diversity and abundance of pollinator species in both island groups, including two avian endemics, the Seychelles Sunbird and the Seychelles Bulbul, and, in Mauritius, the Mauritius Gray White-eye, the Red Fody and the Red-whiskered Bulbul. Importantly, the latter two species were encountered only at the restored site, not in the unrestored area which was dominated by non-native plants.

What these kinds of studies are showing us is that, far from being 'fragile', nature is in fact quite robust. Up to a point it can recover from human-led assaults on its integrity and bounce back like a prize fighter, perhaps a bit bruised and battered but nonetheless ready to take on all comers. Until of course the knockout blow is landed and the boxer (or ecosystem) is punished beyond all restitution. That can happen, though it takes a lot to destroy an ecosystem completely – and before it occurs, interventions such as allowing vegetation to reestablish can be extremely important.

This is even true for tropical forests, which can recover from disturbances such as logging in as little as 20 years if they're allowed to simply get on with it. A 2021 study of 77 sites showed that natural regrowth of tropical forests is faster and more efficient than tree planting for restoring habitats. It goes without saying that habitat restoration is important for storing carbon in an effort to reduce the effects of climate change. It is one of many elements proposed by Project Drawdown, and their 'Table of Solutions' ranks tropical forest restoration in the top 5–10 ways of reducing CO_2 in the atmosphere.

Globally, some very high-profile forest restoration schemes are currently under way. These include Grain for Green, China's attempt to restore vegetation to abandoned farmland to reduce soil erosion and flooding. This has so far involved more than 100 million people and converted tens of millions of hectares to natural vegetation. Then there's the Great Green Wall, a multinational initiative in Africa aimed at restoring the vegetation on the southern edge of the Sahara to combat desertification and mitigate climate change. Several countries – for example, India and Ethiopia – have also made a great deal of noise about marshalling huge public efforts to plant hundreds of millions of trees in a single day.

These big schemes generate a lot of publicity for actions on climate change and are making a difference for regional ecology. Planting trees and allowing habitats to restore themselves, however, is not enough to lessen the effects of climate change, because there will never be enough trees in the world to reduce CO_2 to pre-industrial levels. That's not to say that habitat restoration isn't important: it is, it's just not a silver-bullet solution to climate change, rather part of our toolbox of things that we can do. Just as importantly, restoring habitats provides more opportunities for species to move in response to changing climates, and to recolonise areas from where they have been extirpated. Functioning ecosystems, if you need reminding, support not just wildlife but also human societies, in ways both tangible and unquantifiable.

Of course, in the long term, without pollinators, including birds, and seed dispersers (especially birds), restored forests will not flourish. Let's not forget: species do not occur in isolation, and the biodiversity of species interactions is fundamental to the ecology of the planet. It's important to think about what we're aiming for in large-scale conservation projects, however, and to consider the likely end point of restoration. This can change over time due to a phenomenon that ecologists call the 'shifting baseline' – and to understand that, let's consider a culinary analogy.

Why conservation of habitats is like paella

In April and May 2018 I led field work on Tenerife, initially with students, then joined by colleagues Pablo Gorostiague, Cala Castellanos and Chris Mackin, plus my wife Karin, all of whom you have met earlier in this book. Our aim was to collect more data on the invasive Tree Tobacco. After our students and colleagues had left, Karin and I continued the field work, and one day, tired and hungry after sweating our way through the Malpais de Güímar counting and measuring plants, we ended up at a small seaside restaurant – where we ordered the local paella. I posted a picture on Facebook of Karin tucking into a plate piled high with the Spanish specialty, with a comment that 'after a hot day of collecting data there's nothing better than a nice big Tenerife paella!'

A Spanish colleague spotted the image and commented that the dish was 'closer to being an *arroz con cosas* than a paella'. The term translates as 'rice with things' and is used to convey the fact that the original Valencian dish of paella has been bastardised and changed across the Spanish-speaking world, and no longer reflects its culinary tradition. At the time I knew little of that culinary tradition, so I looked at the Wikipedia entry for paella. It makes for interesting reading, not least for the fact that in the original dish one of the main ingredients was the meat of Southern Water Voles cooked on an open fire fuelled by wood from orange and pine trees to give a distinctive smoky flavour. There was also a lot of geographic variation, however, with different parts of Spain constructing the dish according to locally available foods. So what constitutes an 'authentic' paella is highly debatable. Not only that, but versions of the dish have appeared across the world, in Latin America, the Caribbean, southern USA and the Philippines (where it's called 'paelya'). The only constant culinary ingredient is rice; the rest is down to local tastes.

Although there was no sign of rodent flesh or naked flames in the paella that we ate, it was certainly delicious. But the comment about *arroz con cosas* suggested to me that there are analogous shifting baselines both in cooking and in conservation.

The idea of a shifting baseline is that expectations of what is 'correct' or 'normal' or 'natural' change over time depending upon what each generation has experienced. It's been applied mainly in conservation. For example, the Lake District of England is seen by many as a 'natural' landscape of rolling hills and low mountains, but originally it would have been covered in deciduous forest. Likewise large parts of Tenerife contain a high proportion of alien plants (such as agave, prickly pear and the Tree Tobacco), but local people and visitors alike see this as natural. During one of my Tenerife field courses, a waiter in a restaurant shouted at a student because she was carrying a small piece from an agave that had broken off along a trail that we had recently walked. I intervened and showed him our field-work permit, explaining that agaves are not native to the Canary Islands. But he was not happy that we were, in his view, despoiling the natural environment into which he had been born.

Perhaps more relevant for a book about bird–flower interactions, we know that large areas of Central and South America that were once considered untouched, 'virgin' rainforest, were in fact heavily populated urban centres in pre-colonial times. Thus, the ecological context in which the ancient inhabitants developed cultural relationships with hummingbirds has almost certainly changed since European conquest of these areas. But the indigenous peoples of the Americas undoubtedly also shifted the baseline from what was there prior to their arrival from Asia thousands of years earlier.

Concepts of what is 'natural' thus change over time. Returning a landscape to an original condition would mean a drastic shift in the composition of the vegetation. But what point do we return it to? One hundred years ago? One thousand? Ten thousand? It's an issue that is widely debated in the conservation literature, especially in relation to rewilding.

The use of paella as an analogy is apt because, just like landscapes, over time this dish has changed as it has been adapted by different chefs. What is currently served in restaurants only partially reflects how it was originally cooked, but we perceive this as what paella 'should' look and taste like. Other than for epicurean purists, our culinary expectations have changed because there's been a shift in the paella baseline.

Regardless of what our expectations or goals might be, in order to achieve success in this decade of ecosystem restoration we need knowledge and evidence, information hard won from on-the-ground studies.[1] The destruction and degradation of habitats, as we have seen, are the greatest threats to bird-pollinated plants and the birds on which they depend. But now, overlying this full-frontal assault on the natural world, is a (relatively) new kid on the block. It's a bully that's been hanging around for a couple of decades, making threats about what it's capable of that have been mainly ignored as idle or exaggerated, except by a few lone specialist voices. Now that bully has grown up into a violent adult capable of unspeakable and deadly damage. Climate change has come of age, and every week it seems that we are faced with news reports of drought, flooding, wildfires, or violent hurricanes and typhoons.

Twenty or thirty years ago I admit that I was one of those who thought that climate change was something which was going to happen in the distant future. Although we were seeing changes such as the timing of flowering in spring plants, many of us thought that

---

1 Though, increasingly, technologies such as satellite and drone imagery, LIDAR, eDNA and AI are contributing to the effort.

catastrophic climate change, if it occurred, was going to happen after I and my children were gone, by which time we would have solutions to this issue. But we've run out of time, and the scale of the problems that face us was brought sharply into focus during that austral summer of 2019/20. The Australian news channels were reporting constantly on the fires, their impacts on people and wildlife, and the political fallout for a government that seemed to be deaf and blind to how what was happening was linked to worldwide climatic shifts. As I write in 2023 the news is full of stories about epic floods and wildfires, droughts and deluges. It's no coincidence that globally this year is the hottest on record, with record-breaking sea temperatures.

Scientists such as myself and others you've encountered in the book have been using the knowledge of both past and future possible climates to try to understand how the relationships between birds and flowers might change in the future. From work that I did with Bo Dalsgaard, André Rech and others we know that relationships between plants and their pollinators, including birds, are highly sensitive to climate. It goes without saying that as weather patterns change and regions become wetter or drier, warmer or colder, we can expect to see shifts in the distributions of species as they move or are lost from an area. There will be winners and losers.

Other researchers have built on this work and tried to finesse our understanding of how climate change might affect these long-standing interactions between hummingbirds and their nectar sources. A recent example is the work of Daniela Remolina-Figueroa and colleagues from Mexico, who asked whether the relationships between 12 endemic Mexican hummingbirds and 118 of their nectar plants would be 'together forever'. Using an approach called ecological niche modelling, they estimated that by the 2040s–2080s, almost half of the hummingbirds and plants will have smaller ranges because the climate will be unsuitable for them, resulting in a mismatch of the occurrence of birds and their nectar sources. This in turn will lead to new assemblages of birds and flowers that previously did not interact. Again, it emphasises the importance of having enough habitat across a landscape for species to move into when required, but 'how much habitat' and 'where to locate it' are still unclear.

More questions than birds, more puzzles than plants

As an ecological scientist I sometimes think that we are really just tourists. We might kid ourselves that a career spent observing nature and performing experiments gives us special insights into how the Earth's communities and ecosystems function. But like the holidaymaker who spends a summer fortnight 'getting to know the locals', we are really only seeing a small fraction of life. We might have a few profound insights about conditions that are restricted in time and space, but how do we test these over the millennia across which many processes function, and the thousands of square kilometres of this planet that are still not studied in any depth? Not only that, but the more I learn, the more I observe and the more that I try to find answers, the more I realise how little I truly know.

Birds are arguably the best-studied group of organisms on the planet: whole libraries have been written about their taxonomy, biology, behaviour, migration and evolutionary history. Flowering plants likewise have held the attention of botanists for hundreds of years, and the living herbarium and library collections of institutions such as the Royal Botanic Gardens, Kew are truly inspiring accumulations of knowledge and data. Yet when we bring together these two fields – ornithology and botany, adding a portion of ecology and a sprinkle of palaeontology to spice up the dish – suddenly we realise just how much there is to discover about the biodiversity of interactions between even familiar sets of organisms.

I wrote those last two paragraphs in August 2022, sitting outside the tent that Karin and I shared at the Mpala Research Centre River Camp in Kenya. We were both recovering from COVID infections that had laid us low for a week but not dimmed our curiosity or enjoyment of that wonderful country. It was almost 6 p.m., and the tropical sun was quickly descending, shadows were lengthening, and the savanna was changing colour by the second. Just below our camp the Ewaso Ny'iro River was flowing ponderously towards a bend in the river, beyond which I could not see. What was here a thousand years ago? Or a million years ago? What will it be like in ten thousand years' time, or even next year if the present drought ends and the rains return? Which plants will be favoured by the attentions of the birds, and which birds will benefit from the bounties of the flowers? The chances were that I wouldn't be there to see it, but I took comfort in knowing that in the surrounding tents, and out in the field, were the highly committed African and European students who form part of the next generation of ecological scientists and conservationists. Long may they continue their work to understand and conserve birds and flowers.

Species names

African Monarch	*Danaus chrysippus*
African Sacred Ibis	*Threskiornis aethiopicus*
African Tulip Tree	*Spathodea campanulata*
Amani Sunbird	*Hedydipna pallidigaster*
American Mistletoe	*Phoradendron leucarpum*
Amethyst Sunbird	*Chalcomitra amethystina*
Anna's Hummingbird	*Calypte anna*
Antillean Crested Hummingbird	*Orthorhyncus cristatus*
Ashy Flowerpecker	*Dicaeum vulneratum*
Australian White Ibis	*Threskiornis molucca*
Bald Eagle	*Haliaeetus leucocephalus*
Bananaquit	*Coereba flaveola*
Bee Hummingbird	*Mellisuga helenae*
Bird of Paradise (plant)	*Strelitzia reginae*
Bird's-foot-trefoil	*Lotus corniculatus*
Bismarck Black Myzomela	*Myzomela pammelaena*
Black Currawong	*Strepera fuliginosa*
Black Mamo	*Drepanis funerea*
Black Sunbird	*Leptocoma aspasia*
Black-capped Bulbul	*Rubigula melanicterus*
Black-chinned Hummingbird	*Archilochus alexandri*
Blue-faced Honeyeater	*Entomyzon cyanotis*
Blue-headed Hummingbird	*Riccordia bicolor*
Blue Larkspur	*Delphinium nelsonii*
Blue-mantled Thornbill	*Chalcostigma stanleyi*
Bluethroat	*Luscinia svecica*
Bohemian Waxwing	*Bombycilla garrulus*
Brahminy Starling	*Sturnia pagodarum*
Broad-billed Hummingbird	*Cynanthus latirostris*
Brown Rat	*Rattus norvegicus*
Brown Tree Snake	*Boiga irregularis*
Buff-tailed Coronet	*Boissonneaua flavescens*

Canary Island Bellflower	*Canarina canariensis*
Canary Island Chiffchaff	*Phylloscopus canariensis*
Canary Island Foxglove	*Digitalis (Isoplexis) canariensis*
Cape Glossy Starling	*Lamprotornis nitens*
Cape Sugarbird	*Promerops cafer*
Cape Weaver	*Ploceus capensis*
Cetti's Warbler	*Cettia cetti*
Chatham Island Bellbird	*Anthornis melanocephala*
Cherry Plum	*Prunus cerasifera*
Coal Tit	*Periparus ater*
Coffee Berry Borer	*Hypothenemus hampei*
Common Cactus-Finch	*Geospiza scandens*
Common Chaffinch	*Fringilla coelebs*
Common Chiffchaff	*Phylloscopus collybita*
Common Foxglove	*Digitalis purpurea*
Common Guava	*Psidium guajava*
Common Myna	*Acridotheres tristis*
Common Primrose	*Primula vulgaris*
Common Scale-backed Antbird	*Willisornis poecilinotus*
Common Starling	*Sturnus vulgaris*
Common Sunbird-Asity	*Neodrepanis coruscans*
Common Wood-Pigeon	*Columba palumbus*
Common Yellowthroat	*Geothlypis trichas*
Copper-throated Sunbird	*Leptocoma calcostetha*
Costa's Hummingbird	*Calypte costae*
Coyote Tobacco	*Nicotiana attenuata*
Crested Auklet	*Aethia cristatella*
Crested Berrypecker	*Paramythia montium*
Crimson Sunbird	*Aethopyga siparaja*
Crow Honeyeater	*Gymnomyza aubryana*
Cuban Emerald	*Riccordia ricordii*
Cuban Green Woodpecker	*Xiphidiopicus percussus*
Darjeeling Woodpecker	*Dendrocopos darjellensis*
Dark-billed Cuckoo	*Coccyzus melacoryphus*
Dusky Sunbird	*Cinnyris fuscus*
Eastern Olive Sunbird	*Cyanomitra olivacea*
Eastern Spinebill	*Acanthorhynchus tenuirostris*
Egyptian Vulture	*Neophron percnopterus*
Elegant Sunbird	*Aethopyga duyvenbodei*
Eurasian Blackbird	*Turdus merula*

Eurasian Blackcap	*Sylvia atricapilla*
Eurasian Blue Tit	*Cyanistes caeruleus*
Eurasian Collared-Dove	*Streptopelia decaocto*
European Mistletoe	*Viscum album*
Feral Pigeon	*Columba livia*
Fire-breasted Flowerpecker	*Dicaeum ignipectus*
Fire-tailed Myzornis	*Myzornis pyrrhoura*
Fire-tailed Sunbird	*Aethopyga ignicauda*
Flame of the Forest	*Delonix regia*
Fork-tailed Sunbird	*Aethopyga christinae*
Galápagos Dove	*Zenaida galapagoensis*
Galápagos Flycatcher	*Myiarchus magnirostris*
Garden Warbler	*Sylvia borin*
Geiger Tree	*Cordia sebestena*
Giant Hummingbird	*Patagona gigas*
Gila Woodpecker	*Melanerpes uropygialis*
Glittering-bellied Emerald	*Chlorostilbon lucidus*
Glittering-throated Emerald	*Chionomesa fimbriata*
Goldcrest	*Regulus regulus*
Golden Camellia	*Camellia petelotii*
Golden Wattle	*Acacia pycnantha*
Golden-winged Sunbird	*Drepanorhynchus reichenowi*
Goliath Tarantula	*Theraphosa blondi*
Gran Canaria Blue Chaffinch	*Fringilla polatzeki*
Gray-hooded Sierra Finch	*Phrygilus gayi*
Green-backed Firecrown	*Sephanoides sephaniodes*
Green-backed Honeyeater	*Glycichaera fallax*
Green-throated Carib	*Eulampis holosericeus*
Green Woodhoopoe	*Phoeniculus purpureus*
Guaba	*Inga edulis*
Hawaii Akepa	*Loxops coccineus*
Hawaii Amakihi	*Chlorodrepanis virens*
Herring Gull	*Larus argentatus*
Holly	*Ilex aquifolium*
House Sparrow	*Passer domesticus*
Iiwi	*Drepanis coccinea*
Imperial Fritillary	*Fritillaria imperialis*
Indian Peafowl	*Pavo cristatus*
Indian Pied Myna (or Starling)	*Gracupica contra*
Island Canary	*Serinus canaria*

Ivory-billed Woodpecker	*Campephilus principalis*
Javan Sunbird	*Aethopyga mystacalis*
Juan Fernandez Firecrown	*Sephanoides fernandensis*
Kauai Oo	*Moho braccatus*
Laughing Dove	*Spilopelia senegalensis*
Least Seedsnipe	*Thinocorus rumicivorus*
Lesser Honeyguide	*Indicator minor*
Light-vented Bulbul	*Pycnonotus sinensis*
Little Egret	*Egretta garzetta*
Little Spiderhunter	*Arachnothera longirostra*
Long-bearded Melidectes	*Melidectes princeps*
Long-tailed Hermit	*Phaethornis superciliosus*
Loveridge's Sunbird	*Cinnyris loveridgei*
Malachite Sunbird	*Nectarinia famosa*
Mariqua Sunbird	*Cinnyris mariquensis*
Masked Flowerpiercer	*Diglossa cyanea*
Mauritius Gray White-eye	*Zosterops mauritianus*
Medium Ground-Finch	*Geospiza fortis*
Monarch	*Danaus plexippus*
Moorland Chat	*Pinarochroa sordida*
Mount Teide Bugloss	*Echium wildpretii*
Mountain Sunbird	*Aethopyga jefferyi*
Mute Swan	*Cygnus olor*
New Holland Honeyeater	*Phylidonyris novaehollandiae*
New Zealand Bellbird	*Anthornis melanura*
New Zealand Kaka	*Nestor meridionalis*
New Zealand Pigeon	*Hemiphaga novaeseelandiae*
Nile Valley Sunbird	*Hedydipna metallica*
Noisy Miner	*Manorina melanocephala*
Ocotillo	*Fouquieria splendens*
Olive-backed Sunbird	*Cinnyris jugularis*
Orange-breasted Sunbird	*Anthobaphes violacea*
Painted Honeyeater	*Grantiella picta*
Palestine Sunbird	*Cinnyris osea*
Passenger Pigeon	*Ectopistes migratorius*
Poinsettia	*Euphorbia pulcherrima*
Purple Sunbird	*Cinnyris asiaticus*
Purple-rumped Sunbird	*Leptocoma zeylonica*
Purple-throated Carib	*Eulampis jugularis*
Rainbow Lorikeet	*Trichoglossus moluccanus*

Red Fody	*Foudia madagascariensis*
Red-legged Honeycreeper	*Cyanerpes cyaneus*
Red-rumped Cacique	*Cacicus haemorrhous*
Red-whiskered Bulbul	*Pycnonotus jocosus*
Regent Honeyeater	*Anthochaera phrygia*
Rose-ringed Parakeet	*Psittacula krameri*
Royal Poinciana	*Delonix regia*
Ruby-crowned Tanager	*Tachyphonus coronatus*
Ruby-throated Hummingbird	*Archilochus colubris*
Rufous Hummingbird	*Selasphorus rufus*
Rufous-breasted Hermit	*Glaucis hirsutus*
Rufous-collared Sparrow	*Zonotrichia capensis*
Saguaro	*Carnegiea gigantea*
Sangihe White-eye	*Zosterops nehrkorni*
Santa Marta Sabrewing	*Campylopterus phainopeplus*
Sao Tome Sunbird	*Dreptes thomensis*
Sardinian Warbler	*Curruca melanocephala*
Scarlet Gilia	*Ipomopsis aggregata*
Scarlet Honeyeater	*Myzomela sanguinolenta*
Scarlet-chested Sunbird	*Chalcomitra senegalensis*
Scintillant Hummingbird	*Selasphorus scintilla*
Sclater's Myzomela	*Myzomela sclateri*
Seychelles Bulbul	*Hypsipetes crassirostris*
Seychelles Sunbird	*Cinnyris dussumieri*
Shining Sunbeam	*Aglaeactis cupripennis*
Shining Sunbird	*Cinnyris habessinicus*
Shiny Cowbird	*Molothrus bonariensis*
Silvereye	*Zosterops lateralis*
Souimanga Sunbird	*Cinnyris sovimanga*
South American Bushmaster	*Lachesis muta*
Southern Double-collared Sunbird	*Cinnyris chalybeus*
Southern Water Vole	*Arvicola sapidus*
Spanish Sparrow	*Passer hispaniolensis*
Speckled Mousebird	*Colius striatus*
Spotted Snapweed	*Impatiens balsamina*
Sunshine Wattle	*Acacia terminalis*
Swinhoe's White-eye	*Zosterops simplex*
Sword-billed Hummingbird	*Ensifera ensifera*
Tacazze Sunbird	*Nectarinia tacazze*
Tayra	*Eira barbara*

Tenerife Blue Chaffinch	*Fringilla teydea*
Tennessee Warbler	*Leiothlypis peregrina*
Tit Berrypecker	*Oreocharis arfaki*
Tree Bumblebee	*Bombus hypnorum*
Tree Tobacco	*Nicotiana glauca*
Trumpet Vine	*Campsis radicans*
Tui	*Prosthemadera novaeseelandiae*
Turquoise-throated Puffleg	*Eriocnemis godini*
Tyrian Metaltail	*Metallura tyrianthina*
Variable Oriole	*Icterus pyrrhopterus*
Variable Sunbird	*Cinnyris venustus*
Verdin	*Auriparus flaviceps*
Vervet Monkey	*Chlorocebus pygerythrus*
Warbling White-eye	*Zosterops japonicus*
Western Honey Bee	*Apis mellifera*
Western Violet-backed Sunbird	*Anthreptes longuemarei*
White-chested White-eye	*Zosterops albogularis*
White-tipped Sicklebill	*Eutoxeres aquila*
White-winged Dove	*Zenaida asiatica*
Wild Chilli	*Capsicum frutescens*
Woolly Mammoth	*Mammuthus primigenius*
Yellow Fever Tree	*Acacia (Vachellia) xanthophloea*
Yellow Rattle	*Rhinanthus minor*
Yellow Warbler	*Setophaga petechia*
Yellow-bellied Sunbird-Asity	*Neodrepanis hypoxantha*
Yellow-breasted Chat	*Icteria virens*
Yunnan Banana	*Musa yunnanensis*

Sources and further reading

Introduction – Encounters with birds and flowers

Abrahamczyk, S. (2019) Comparison of the ecology and evolution of plants with a generalist bird pollination system between continents and islands worldwide. *Biological Reviews* 94: 1658–1671.

Anderson, S., Kelly, D., Robertson, A. and Ladley, J. (2016) Pollination by birds: a functional evaluation. In Sekercioglu, Ç., Wenny, D. and Whelan, C. (eds), *Why Birds Matter: Avian Ecological Function and Ecosystem Services*. University of Chicago Press, Chicago, IL, pp. 73–106.

Buchmann, S. and Nabhan, G.P. (1996) *The Forgotten Pollinators*. Island Press, Washington, DC.

Cronk, Q. and Ojeda, I. (2008) Bird-pollinated flowers in an evolutionary and molecular context. *Journal of Experimental Botany* 59: 715–727.

Leimberger, K.G., Dalsgaard, B., Tobias, J.A., Wolf, C. and Betts, M.G. (2022) The evolution, ecology, and conservation of hummingbirds and their interactions with flowering plants. *Biological Reviews* 97: 923–959.

Ollerton, J. (2021) *Pollinators & Pollination: Nature and Society*. Pelagic Publishing, Exeter.

Pauw, A. (2019) A bird's-eye view of pollination: biotic interactions as drivers of adaptation and community change. *Annual Review of Ecology, Evolution, and Systematics* 50: 477–502.

Ratto, F., Simmons, B.I., Spake, R. *et al.* (2018) Global importance of vertebrate pollinators for plant reproductive success: a meta-analysis. *Frontiers in Ecology and the Environment* 16: 82–90. https://doi.org/10.1002/fee.1763

Chapter 1 – Origins of a partnership

Bestwick, J., Unwin, D.M., Butler, R.J., Henderson, D.M. and Purnell, M.A. (2018) Pterosaur dietary hypotheses: a review of ideas and approaches. *Biological Reviews* 93: 2021–2048. https://doi.org/10.1111/brv.12431

Bochenski, Z.M., Tomek, T., Wertz, K. and Świdnicka, E. (2013) The third nearly complete passerine bird from the early Oligocene of Europe. *Journal of Ornithology* 154: 923–931.

Boles, W.E. (2005) Fossil honeyeaters (Meliphagidae) from the Late Tertiary of Riversleigh, north-western Queensland. *Emu – Austral Ornithology* 105: 21–26.

Farlow, J.O. (1987) Speculations about the diet and digestive physiology of herbivorous dinosaurs. *Paleobiology* 13: 60–72.

Goulson, D. (2021) *Silent Earth*. Harper, London.

Han, L., Zhao, Y., Zhao, M., Sun, J., Sun, B. and Wang, X. (2023) New fossil evidence suggests that angiosperms flourished in the Middle Jurassic. *Life* 13: 819. https://doi.org/10.3390/life13030819

Hertel, F. (1995) Ecomorphological indicators of feeding behavior in Recent and fossil raptors. *The Auk* 112: 890–903.

Mayr, G. (2007) New specimens of the early Oligocene Old World hummingbird *Eurotrochilus inexpectatus*. *Journal of Ornithology* 148: 105–111.

Mayr, G. and Micklich, N. (2010) New specimens of the avian taxa *Eurotrochilus* (Trochilidae) and *Palaeotodus* (Todidae) from the early Oligocene of Germany. *Paläontologische Zeitschrift* 84: 387–395.

Mayr, G. and Wilde, V. (2014) Eocene fossil is earliest evidence of flower-visiting by birds. *Biology Letters* 10: 20140223.

Olesen, J.M. and Valido, A. (2003) Lizards as pollinators and seed dispersers: an island phenomenon. *Trends in Ecology and Evolution* 18: 177–181.

Ollerton, J. (2021) *Pollinators & Pollination: Nature and Society.* Pelagic Publishing, Exeter.

Ollerton, J. and Coulthard, E. (2009) Evolution of animal pollination. *Science* 326: 808–809.

Ollerton, J., Tarrant, S. and Winfree, R. (2011) How many flowering plants are pollinated by animals? *Oikos* 120: 321–326.

Pittman M., Barlow L.A., Kaye, T.G. and Habib, M.B. (2021) Pterosaurs evolved a muscular wing–body junction providing multifaceted flight performance benefits: Advanced aerodynamic smoothing, sophisticated wing root control, and wing force generation. *Proceedings of the National Academy of Sciences of the USA* 118: e2107631118. https://doi.org/10.1073/pnas.2107631118

Rodger, J., Bennett, J.M., Razanajatovo, M. *et al.* (2021) Widespread vulnerability of flowering plant seed production to pollinator declines. *Science Advances* 7: eabd3524. https://doi.org/10.1126/sciadv.abd3524

van der Kooi, C.J. and Ollerton, J. (2020) The origins of flowering plants and pollinators. *Science* 368: 1306–1308.

Wappler, T. and Engel, M. (2003) The Middle Eocene bee faunas of Eckfeld and Messel, Germany (Hymenoptera: Apoidea). *Journal of Paleontology* 77: 908–921.

Waser, N.M., Chittka, L., Price, M.V., Williams, N. and Ollerton, J. (1996) Generalization in pollination systems, and why it matters. *Ecology* 77: 1043–1060.

Wester, P. (2019) First observations of nectar-drinking lizards on the African mainland. *Plant Ecology and Evolution* 152: 78–83. https://doi.org/10.5091/plecevo.2019.1513

Wu, Y., Ge, Y., Hu, H. *et al.* (2023) Intra-gastric phytoliths provide evidence for folivory in basal avialans of the Early Cretaceous Jehol Biota. *Nature Communications* 14, 4558. https://doi.org/10.1038/s41467-023-40311-z. See also: https://scitechdaily.com/120-million-year-old-fossil-unveils-birds-leaf-eating-past

Chapter 2 – Surprising variety

Abrahamczyk, S. (2019) Comparison of the ecology and evolution of plants with a generalist bird pollination system between continents and islands worldwide. *Biological Reviews* 94: 1658–1671.

Anderson, B., Cole, W.W. and Barrett, S.C.H. (2005) Specialized bird perch aids cross-pollination. *Nature* 435: 41–42.

Barrowclough, G.F., Cracraft, J., Klicka, J. and Zink, R.M. (2016) How many kinds of birds are there and why does it matter? *PLoS One* 11: e0166307. https://doi.org/10.1371/journal.pone.0166307

Bartoš, M. and Janeček, Š. (2014) Pollinator-induced twisting of flowers sidesteps floral architecture constraints. *Current Biology* 24: R793–R795.

Birds of the World (2022) Cornell Laboratory of Ornithology, Ithaca, NY. https://birdsoftheworld.org/bow

Cheke, R. and Mann, C. (2020) Western Violet-backed Sunbird (*Anthreptes longuemarei*), version 1.0. *Birds of the World*. Cornell Lab of Ornithology, Ithaca, NY. https://doi.org/10.2173/bow.wvbsun1.01

de Waal, C., Anderson B. and Barrett, S.C.H. (2012) The natural history of pollination and mating in bird-pollinated *Babiana* (Iridaceae). *Annals of Botany* 109: 667–679.

Fleming, T.H. and Muchhala, N. (2008) Nectar-feeding bird and bat niches in two worlds: pantropical comparisons of vertebrate pollination systems. *Journal of Biogeography* 35: 764–780.

Geerts, S. and Pauw, A. (2009) African sunbirds hover to pollinate an invasive hummingbird-pollinated plant. *Oikos* 118: 573–579.

Gill, F., Donsker, D. and Rasmussen, P. (eds) (2023) *IOC World Bird List* (v13.2). www.worldbirdnames.org

Janeček, Š., Patáčová, E., Bartoš, M., Padyšáková, E., Spitzer, L. and Tropek, R. (2011) Hovering sunbirds in the Old World: occasional behaviour or evolutionary trend? *Oikos* 120: 178–183.

Janeček, Š., Bartoš, M. and Njabo, K.Y. (2015) Convergent evolution of sunbird pollination systems of *Impatiens* species in tropical Africa and hummingbird systems of the New World. *Biological Journal of the Linnean Society* 115: 127–133.

Kiepiel, I., Brown, M. and Johnson, S.D. (2022) A generalized bird pollination system in *Schotia brachypetala* (Fabaceae). *Plant Biology* 24: 806–814.

Miller, R.S. (1985) Why hummingbirds hover. *The Auk* 102: 722–726.

Ollerton, J., Koju, N.P., Maharjan, S.R. and Bashyal, B. (2019) Interactions between birds and flowers of *Rhododendron* spp., and their implications for mountain communities in Nepal. *Plants, People, Planet* 2: 320–325. https://doi.org/10.1002/ppp3.10091

Osipova, E., Barsacchi, R., Brown, T. *et al.* (2023) Loss of a gluconeogenic muscle enzyme contributed to adaptive metabolic traits in hummingbirds. *Science* 379: 185–190. https://doi.org/10.1126/science.abn7050. See also: https://phys.org/news/2023-01-hummingbirds-flight-evolved-lost-gene.html

Pyke, G.H. (1981) Why hummingbirds hover and honeyeaters perch. *Animal Behaviour* 29: 861–867.

Regan, E.C., Santini, L., Ingwall-King, L. *et al.* (2015) Global trends in the status of bird and mammal pollinators. *Conservation Letters* 8: 397–403.

Reilly, J. (2019) *The Ascent of Birds*. Pelagic Publishing, Exeter.

Sachatamia Lodge Webcam (2023) www.youtube.com/watch?v=1ZUHgXEGgic

Warrick, D., Tobalske, B. and Powers, D. (2005) Aerodynamics of the hovering hummingbird. *Nature* 435: 1094–1097. https://doi.org/10.1038/nature03647

Wells, D.J. (1993) Ecological correlates of hovering flight of hummingbirds. *Journal of Experimental Biology* 178: 59–70.

Wester, P. (2013) Feeding on the wing: hovering in nectar-drinking Old World birds. More common than expected. *Emu – Austral Ornithology* 114: 171–183.

Chapter 3 – Keeping it in the family

Alcorn, S.M., McGregor, S.E. and Olin, G. (1961) Pollination of saguaro cactus by doves, nectar-feeding bats, and honey bees. *Science* 133: 1594–1595.

Bartley, G. and Swash, A. (2022) *Hummingbirds: A Celebration of Nature's Jewels*. Princeton University Press, Princeton, NJ.

Cheke R.A., Mann, C.F. and Allen, R. (2001) *Sunbirds: A Guide to the Sunbirds, Spiderhunters, Sugarbirds and Flowerpeckers of the World*. Yale University Press, New Haven, CT.

Cotton, P.A. (2001) The behavior and interactions of birds visiting *Erythrina fusca* flowers in the Colombian Amazon. *Biotropica* 33: 662–669.

Coyle, C.M. (2022) Songbirds as Pollinators project. https://carolyncoyle.wixsite.com/sapproject

Coyle, C.M. and Gannon, D.G. (2021) Observations of orange-crowned warblers in vine maple. *Northwestern Naturalist* 102: 94–95.

Craig, A.J.F. and Feare, C.J. (2020) Brahminy Starling (*Sturnia pagodarum*), version 1.0. *Birds of the World*. Cornell Lab of Ornithology, Ithaca, NY. https://doi.org/10.2173/bow.brasta1.01

Cuta-Pineda, J.A., Arias-Sosa, L.A. and Pelayo, R.C. The flowerpiercers interactions with a community of high Andean plants. *Avian Research* 12: 22. https://doi.org/10.1186/s40657-021-00256-7

Dalsgaard, B. (2011) Nectar-feeding and pollination by the Cuban Green Woodpecker (*Xiphidiopicus percussus*) in the West Indies. *Ornitologia Neotropical* 22: 447–451.

da Silva, L.P., Ramos, J.A., Olesen, J.M., Traveset, A. and Heleno, R.H. (2014) Flower visitation by birds in Europe. *Oikos* 123: 1377–1383.

Drezner, T.D. (2014) The keystone saguaro (*Carnegiea gigantea*, Cactaceae): a review of its ecology, associations, reproduction, limits, and demographics. *Plant Ecology* 215: 581–595.

Fitzsimons, J.A. (2019) Observations of nectarivory in the Little Raven (*Corvus mellori*) and a review of nectarivory in other *Corvus* species. *Wilson Journal of Ornithology* 131: 382–386.

Ford, H.A. (1985) Nectarivory and pollination by birds in southern Australia and Europe. *Oikos* 44: 127–131.

Franklin, D.C. (2007) Sulphur-crested Cockatoo *Cacatua galerita* feeds on nectar. *Northern Territory Naturalist* 19: 46–47.

O'Donnell, C.F.J. and Dilks P.J. (1989) Feeding on fruits and flowers by insectivorous forest birds. *Notornis* 36: 72–76.

O'Donnell, C.F.J. and Dilks P.J. (1994) Foods and foraging of forest birds in temperate rain-forest, south Westland, New Zealand. *New Zealand Journal of Ecology* 18: 87–107.

O'Donnell, C.F.J., Reardon, J.T. and Hoare, J.M. (2011) Nectar feeding by rock wren (*Xenicus gilviventris*). *Notornis* 58: 46–47.

Fleming, T.H., Tuttle, M.D. and Horner, M.A. (1996) Pollination biology and the relative importance of nocturnal and diurnal pollinators in three species of Sonoran Desert columnar cacti. *Southwestern Naturalist* 41: 257–269.

Fleming, T.H., Sahley, C.T., Holland, J.N., Nason, J.D. and Hamrick, J.L. (2001) Sonoran Desert columnar cacti and the evolution of generalized pollination systems. *Ecological Monographs* 71: 511–530.

Li, Z., Wang, M., Stidham, T.A., Zhou, Z. and Clarke, J. (2022) Novel evolution of a hyper-elongated tongue in a Cretaceous enantiornithine from China and the evolution of the hyolingual apparatus and feeding in birds. *Journal*

of Anatomy 240: 627–638. https://doi.org/10.1111/joa.13588. See also: www.sci-news.com/paleontology/brevirostruavis-macrohyoideus-10367.html
Medrano, F., Sepúlveda, M.J., Heynen, I., Boesman, P.F.D. and Kirwan, G.M. (2022) Giant Hummingbird (*Patagona gigas*), version 2.0. *Birds of the World*. Cornell Lab of Ornithology, Ithaca, NY. https://doi.org/10.2173/bow.giahum1.02
Nicol, S.C., McLachlan-Troup, T.A. and McQuillan, P.B. (2023) Pollination of *Richea dracophylla* (Ericaceae) and the role of the nocturnal ant *Notoncus hickmani* (Formicinae). *Austral Ecology* 00: 1–17. https://doi.org/10.1111/aec.13421
Ollerton, J., Koju, N.P., Maharjan, S.R. and Bashyal, B. (2019) Interactions between birds and flowers of *Rhododendron* spp., and their implications for mountain communities in Nepal. *Plants, People, Planet* 2: 320–325. https://doi.org/10.1002/ppp3.10091
Ó Marcaigh, F., Kelly, D.J., O'Connell, D.P. *et al.* (2022) Small islands and large biogeographic barriers have driven contrasting speciation patterns in Indo-Pacific sunbirds (Aves: Nectariniidae). *Zoological Journal of the Linnean Society* 198: 72–92. https://doi.org/10.1093/zoolinnean/zlac081. See also: https://phys.org/news/2022-10-beautiful-bird-species-remote-indonesian.html
Paiaro, V., Cocucci, A.A., Oliva, G.E. and Sérsic, A.N. (2017) The role of facultatively nectarivorous birds as pollinators of *Anarthrophyllum desideratum* in the Patagonian steppe: a geographical approach. *Ecología Austral* 27: 312–325.
Rand, A.L. (1967) The flower-adapted tongue of a Timaliinae bird and its implications. *Fieldiana Zoology* 51: 53–63.
Rocca, M.A. and Sazima, M. (2010) Beyond hummingbird-flowers: the other side of ornithophily in the Neotropics. *Oecologia Australis* 14: 67–99.
Symes, C.T., Nicolson, S.W. and McKechnie, A.E. (2008) Response of avian nectarivores to the flowering of *Aloe marlothii*: a nectar oasis during dry South African winters. *Journal of Ornithology* 149: 13–22.
Young, T.P. (1982) Bird visitation, seed set, and germination rates in two *Lobelia* species on Mount Kenya. *Ecology* 68: 1983–1986.
Zuccon, D., Pryˆs-Jones, R., Rasmussen, P.C. and Ericson, P.G.P. (2012) The phylogenetic relationships and generic limits of finches (Fringillidae). *Molecular Phylogenetics and Evolution* 62: 581–596.

Chapter 4 – A flower's point of view

Abrahamczyk, S. and Kessler, M. (2015) Morphological and behavioural adaptations to feed on nectar: how feeding ecology determines the diversity and composition of hummingbird assemblages. *Journal of Ornithology* 156: 333–347.
Abrahamczyk, S., Weigend, M., Becker, K. *et al.* (2022) Influence of plant reproductive systems on the evolution of hummingbird pollination. *Ecology and Evolution* 12: e8621. https://doi.org/10.1002/ece3.8621
Anderson, S.H. (2003) The relative importance of birds and insects as pollinators of the New Zealand flora. *New Zealand Journal of Ecology* 27: 83–94.
Castellanos, M.C., Wilson, P. and Thomson, J.D. (2003) Pollen transfer by hummingbirds and bumblebees, and the divergence of pollination modes in *Penstemon*. *Evolution* 57: 2742–2752.

Corlett, R.T. (2004) Flower visitors and pollination in the Oriental (Indomalayan) Region. *Biological Reviews* 79: 497–532. https://doi.org/10.1017/S1464793103006341

Gamba, D. and Muchhala, N. (2023) Pollinator type strongly impacts gene flow within and among plant populations for six Neotropical species. *Ecology* 104: e3845. https://doi.org/10.1002/ecy.3845

Gorostiague P. and Ortega-Baes P. (2017) Pollination biology of *Echinopsis leucantha* (Cactaceae): passerine birds and exotic bees as effective pollinators. *Botany* 95: 53–59.

Gorostiague, P., Ollerton, J. and Ortega-Baes, P. (2023) Latitudinal gradients in biotic interactions: are cacti pollination systems more specialized in the tropics? *Plant Biology* 25: 187–197. https://doi.org/10.1111/plb.13450

Gottsberger, G. and Silberbauer-Gottsberger I. (2006) *Life in the Cerrado: a South American Tropical Seasonal Ecosystem. Vol. I. Origin, Structure, Dynamics and Plant Use*. Reta, Ulm.

Gottsberger, G. and Silberbauer-Gottsberger I. (2006) *Life in the Cerrado: a South American Tropical Seasonal Ecosystem. Vol. II. Pollination and Seed Dispersal.* Reta, Ulm.

Grant V. and Grant K.A. (1965) *Flower Pollination in the Phlox Family*. Columbia University Press, New York.

Keighery, G.J. (1980) Bird pollination in South Western Australia: a checklist. *Plant Systematics and Evolution* 135: 171–176.

Kay, K.M. and Grossenbacher, D.L. (2022) Evolutionary convergence on hummingbird pollination in Neotropical *Costus* provides insight into the causes of pollinator shifts. *New Phytologist* 236: 1572–1583.

Medel, R., López-Aliste, M. and Fontúrbel, F.E. (2022) Hummingbird–plant interactions in Chile: an ecological review of the available evidence. *Avian Research* 13: 100051. https://doi.org/10.1016/j.avrs.2022.100051

Muschett, G. and Fontúrbel, F.E. (2022) A comprehensive catalogue of plant–pollinator interactions for Chile. *Scientific Data* 9: 78. https://doi.org/10.1038/s41597-022-01195-8

Kress, W.J. and Beach, J.H. (1994) Flowering plant reproductive systems. In McDade, L.A., Bawa, K.S., Hespenheide, H. and Hartshorn, G. (eds), *La Selva: Ecology and Natural History of a Neotropical Rain Forest*. University of Chicago Press, Chicago, IL, pp. 161–182.

Machado, I.C. and Lopes, A.V. (2004) Floral traits and pollination systems in the Caatinga, a Brazilian tropical dry forest. *Annals of Botany* 94: 365–376. https://doi.org/10.1093/aob/mch152

Mundi O., Ii, T.A., Chmel, K. *et al.* (2022) The ornithophily of *Impatiens sakeriana* does not guarantee a preference by sunbirds. *Biological Journal of the Linnean Society* 137: 240–249.

Ollerton, J. (2017) Pollinator diversity: distribution, ecological function, and conservation. *Annual Review of Ecology, Evolution, and Systematics* 48: 353–376.

Ollerton, J., Tarrant, S. and Winfree, R. (2011) How many flowering plants are pollinated by animals? *Oikos* 120: 321–326.

Ollerton, J., Liede-Schumann, S., Endress, M E. *et al.* (2019) The diversity and evolution of pollination systems in large plant clades: Apocynaceae as a case study. *Annals of Botany* 123: 311–325.

Rose, J.P. and Sytsma, K.J. (2021) Complex interactions underlie the correlated evolution of floral traits and their association with pollinators in a clade with diverse pollination systems. *Evolution* 75: 1431–1449.

Serrano-Serrano, M.L., Rolland, J., Clark, J.L., Salamin, N. and Perret, M. (2017) Hummingbird pollination and the diversification of angiosperms: an old and successful association in Gesneriaceae. *Proceedings of the Royal Society B* 284: 20162816. https://doi.org/10.1098/rspb.2016.2816

Silberbauer-Gottsberger, I. and Gottsberger, G. (1988) A polinização de plantas do Cerrado. *Revista Brasileira de Biologia* 48: 651–663.

Waser, N.M. and Campbell, D.R. (2004) Ecological speciation in flowering plants. In Dieckmann, U., Doebeli, M., Metz, J.A.J. and Tautz, D. (eds), *Adaptive Speciation*. Cambridge University Press, Cambridge, pp. 264–277.

Wilson, P., Castellanos, M.C., Hogue, J.N., Thomson, J.D. and Armbruster, W.S. (2004) A multivariate search for pollination syndromes among penstemons. *Oikos* 104: 345–361.

Wilson, P., Castellanos, M.C., Wolfe, A.D. and Thomson, J.D. (2006) Shifts between bee and bird pollination in penstemons. In Waser, N.M. and Ollerton, J. (eds), *Plant–Pollinator Interactions: From Specialization to Generalization*. University of Chicago Press, Chicago, IL, pp. 47–68.

Chapter 5 – In the eye of the beholder

Aguilar R., Martén-Rodríguez, S., Avila-Sakar, G. *et al.* (2015) A global review of pollination syndromes: a response to Ollerton *et al.* 2015. *Journal of Pollination Ecology* 17: 126–128.

Aigner, P.A. (2001) Optimality modeling and fitness trade-offs: when should plants become pollinator specialists? *Oikos* 95: 177–184.

Aigner, P.A. (2004) Floral specialization without trade-offs: optimal corolla flare in contrasting pollination environments. *Ecology* 85: 2560–2569.

Aigner, P.A. (2006) The evolution of specialized floral phenotypes in a fine-grained pollination environment. In Waser, N.M. and Ollerton, J. (eds), *Plant–Pollinator Interactions: From Specialization to Generalization*. University of Chicago Press, Chicago, IL, pp. 23–46.

Amorim, M.D., Maruyama, P.K., Baronio, G.J., Azevedo, C.S. and Rech, A.R. (2022) Hummingbird contribution to plant reproduction in the rupestrian grasslands is not defined by pollination syndrome. *Oecologia* 199: 1–12. https://doi.org/10.1007/s00442-021-05103-6

Chmel, K., Ewome, F.L., Gómez, G.U. *et al.* (2021) Bird pollination syndrome is the plant's adaptation to ornithophily, but nectarivorous birds are not so selective. *Oikos* 130: 1411–1424. https://doi.org/10.1111/oik.08052

Dellinger, A.S. (2020) Pollination syndromes in the 21st century: where do we stand and where may we go? *New Phytologist* 228: 1193–1213. https://doi.org/10.1111/nph.16793

Fenster, C.B., Armbruster, W.S., Wilson, P., Dudash, M.R. and Thomson, J.D. (2004) Pollination syndromes and floral specialization. *Annual Review of Ecology, Evolution, and Systematics* 35: 375–403. https://doi.org/10.1146/annurev.ecolsys.34.011802.132347

Ford, H.A. and Forde, N. (1976) Birds as possible pollinators of *Acacia pycnantha*. *Australian Journal of Botany* 24: 793–795. https://doi.org/10.1071/BT9760793

Hingston, A.B. and McQuillan, P.B. (2000) Are pollination syndromes useful predictors of floral visitors in Tasmania? *Austral Ecology* 25: 600–609. https://doi.org/10.1111/j.1442-9993.2000.tb00065.x

Johnson, S.D. and Bond, W.J. (1994) Red flowers and butterfly pollination in the fynbos of South Africa. In Arianoutsou, M. and Groves,

R.H. (eds), *Plant–Animal Interactions in Mediterranean-type Ecosystems. Tasks for Vegetation Science* 31. Springer, Dordrecht. https://doi.org/10.1007/978-94-011-0908-6_13

Knox, R.B., Kenrick, J., Bernhardt, P. *et al.* (1985) Extrafloral nectaries as adaptations for bird pollination in *Acacia terminalis. American Journal of Botany* 72: 1185–1196. https://doi.org/10.1002/j.1537-2197.1985.tb08371.x

Martén-Rodríguez, S., Almarales-Castro, A. and Fenster, C.B. (2009) Evaluation of pollination syndromes in Antillean Gesneriaceae: evidence for bat, hummingbird and generalized flowers. *Journal of Ecology* 97: 348–359. https://doi.org/10.1111/j.1365-2745.2008.01465.x

Mayfield, M.M., Waser, N.M. and Price, M.V. (2001) Exploring the 'most effective pollinator principle' with complex flowers: bumblebees and *Ipomopsis aggregata. Annals of Botany* 88: 591–596. https://doi.org/10.1006/anbo.2001.1500

Ollerton, J., Alarcón, R., Waser, N.M. *et al.* (2009) A global test of the pollination syndrome hypothesis. *Annals of Botany* 103: 1471–1480. https://doi.org/10.1093/aob/mcp031

Ollerton, J. Waser, N.M., Rodrigo Rech, A. and Price, M.V. (2015) Using the literature to test pollination syndromes: some methodological cautions. *Journal of Pollination Ecology* 16: 119–125. https://doi.org/10.26786/1920-7603(2015)17

Sinnott-Armstrong, M.A., Deanna, R., Pretz, C. *et al.* (2022) How to approach the study of syndromes in macroevolution and ecology. *Ecology and Evolution* 12: e8583. https://doi.org/10.1002/ece3.8583

Stanley, D.A. and Cosnett, E. (2021) Catching the thief: nectar robbing behaviour by bumblebees on naturalised *Fuchsia magellanica* in Ireland. *Journal of Pollination Ecology* 29: 240–248.

Stebbins, G.L. (1970) Adaptive radiation of reproductive characteristics in angiosperms I: pollination mechanisms. *Annual Review of Ecology and Systematics* 1: 307–326.

Stone, G.N., Raine, N.E., Prescott, M. and Willmer, P.G. (2003) Pollination ecology of acacias (Fabaceae, Mimosoideae). *Australian Systematic Botany* 16: 103–118.

van der Niet, T., Cozien, R.J., Castañeda-Zárate, M. and Johnson, S.D. (2022) Long-term camera trapping needed to identify sunbird species that pollinate the endangered South African orchid *Satyrium rhodanthum. African Journal of Ecology* 60: 1278– 1282.

Vanstone, V.A. and Paton, D.C. (1988) Extrafloral nectaries and pollination of *Acacia pycnantha* by birds. *Australian Journal of Botany* 36: 519–531.

Waser, N.M., Ollerton, J. and Erhardt, A. (2011) Typology in pollination biology: lessons from an historical critique. *Journal of Pollination Ecology* 3: 1–7.

Waser, N.M., Ollerton, J. and Price, M.V. (2015) Response to Aguilar *et al.*'s (2015) critique of Ollerton *et al.* (2009). *Journal of Pollination Ecology* 17: 1–2.

Waser, N.M., CaraDonna, P.J. and Price, M.V. (2018) Atypical flowers can be as profitable as typical hummingbird flowers. *The American Naturalist* 192: 644–653.

Chapter 6 – Goods and services

Ackerman, J.D., Phillips, R.D., Tremblay, R.L. *et al.* (2023) Beyond the various contrivances by which orchids are pollinated: global patterns in orchid pollination biology. *Botanical Journal of the Linnean Society* 202: 295–324. https://doi.org/10.1093/botlinnean/boac082

Baker, H.G. and Baker, I. (1990) The predictive value of nectar chemistry to the recognition of pollinator types. *Israel Journal of Botany* 39: 157–166.

Brice, A.T., Dahl, K.H. and Grau, C.R. (1989) Pollen digestibility by hummingbirds and psittacines. *The Condor* 91: 681–688.

Coombs, G. and Peter, C.I. (2009) Do floral traits of *Strelitzia reginae* limit nectar theft by sunbirds? *South African Journal of Botany* 75: 751–756.

de Araújo, F.P., Hoffmann, D., Dambolena, J.S., Galetto, L. and Sazima, M. (2021) Nectar characteristics of hummingbird-visited ornithophilous and non-ornithophilous flowers from Cerrado, Brazil. *Plant Systematics and Evolution* 307, 64. https://doi.org/10.1007/s00606-021-01790-z

Dellinger, A.S., Penneys, D.S., Staedler, Y.M., Fragner, L., Weckwerth, W. and Schönenberger, J. (2014) A specialized bird pollination system with a bellows mechanism for pollen transfer and staminal food body rewards. *Current Biology* 24: 1615– 1619.

Dellinger, A.S., Artuso, S., Fernández-Fernández, D.M. and Schönenberger, J. (2021) Stamen dimorphism in bird-pollinated flowers: investigating alternative hypotheses on the evolution of heteranthery. *Evolution* 75: 2589–2599.

del Rio, C.M. (1990) Sugar preferences in hummingbirds: the influence of subtle chemical differences on food choice. *The Condor* 92: 1022–1030.

Filipiak, Z.M., Ollerton, J. and Filipiak, M. (2023) Uncovering the significance of the ratio of food K:Na in bee ecology and evolution. *Ecology* 104: e4110. https://doi.org/10.1002/ecy.4110

Frost, S.K. and Frost, P.G.H. (1981) Sunbird pollination of *Strelitzia nicolai*. *Oecologia* 49: 379–384.

Grant, B.R. (1996) Pollen digestion by Darwin's finches and its importance for early breeding. *Ecology* 77: 489–499.

Hansen, D.M., Olesen, J.M., Mione, T., Johnson, S.D. and Müller, C.B. (2007) Coloured nectar: distribution, ecology, and evolution of an enigmatic floral trait. *Biological Reviews* 82: 83–111.

Johnson, S.D. (1996) Bird pollination in South African species of *Satyrium* (Orchidaceae). *Plant Systematics and Evolution* 203: 91–98.

Johnson, S.D. and Brown, M. (2004) Transfer of pollinaria on birds' feet: a new pollination system in orchids. *Plant Systematics and Evolution* 244: 181–188.

Johnson, S.D. and Nicolson, S.W. (2008) Evolutionary associations between nectar properties and specificity in bird pollination systems. *Biology Letters* 23: 49–52.

Johnson, S.D., Hargreaves, A.L. and Brown, M. (2006) Dark, bitter-tasting nectar functions as a filter of flower visitors in a bird-pollinated plant. *Ecology* 87: 2709–2716.

Karremans, A.P. (2023) *Demystifying Orchid Pollination: Stories of Sex, Lies and Obsession*. Kew Publishing, London.

Liu, Z.J., Chen, L.J., Liu, K.W. *et al.* (2013) Adding perches for cross-pollination ensures the reproduction of a self-incompatible orchid. *PLoS One* 8: e53695. https://doi.org/10.1371/journal.pone.0053695

Magner, E.T., Roy, R., Freund Saxhaug, K. *et al.* (2023) Post-secretory synthesis of a natural analog of iron-gall ink in the black nectar of *Melianthus* spp. *New Phytologist* 239: 2026–2040.

Micheneau, C., Fournel, J. and Pailler, T. (2006) Bird pollination in an angraecoid orchid on Reunion Island (Mascarene Archipelago, Indian Ocean). *Annals of Botany* 97: 965–974.

Nicolson, S.W. (2022) Sweet solutions: nectar chemistry and quality. *Philosophical Transactions of the Royal Society B* 377: 20210163. https://doi.org/10.1098/rstb.2021.0163

Nunes, C.E.P., Castro, M.d.M., Galetto, L. and Sazima, M. (2013) Nectar secretion in *Elleanthus brasiliensis* (Orchidaceae). *Botanical Journal of the Linnean Society* 171: 764–772.

Ollerton, J. (1998) Sunbird surprise for syndromes. *Nature* 394: 726–727.

Ollerton, J. (2006) 'Biological barter': patterns of specialization compared across different mutualisms. In Waser, N.M. and Ollerton, J. (eds), *Plant–Pollinator Interactions: From Specialization to Generalization*. University of Chicago Press, Chicago, IL, pp. 411–435.

Paton, D.C. (1981) The significance of pollen in the diet of the New Holland Honeyeater, *Phylidonyris novaehollandiae* (Aves : Meliphagidae). *Australian Journal of Zoology* 29: 217–224.

Pauw, A. (1998) Pollen transfer on birds' tongues. *Nature* 394: 731–732.

Pyke, G.H. (1978) Optimal foraging in hummingbirds: testing the marginal value theorem. *American Zoologist* 18: 739–752.

Pyke, G.H. and Waser, N.M. (1981) The production of dilute nectars by hummingbird and honeyeater flowers. *Biotropica* 260–270.

Pyke, G.H., Kalman, J.R.M., Bordin, D.M., Blanes, L. and Doble, P.A. (2020) Patterns of floral nectar standing crops allow plants to manipulate their pollinators. *Scientific Reports* 10: 1–10.

Ray, H.A. and Gillett-Kaufman, J.L. (2022) By land and by tree: pollinator taxa diversity of terrestrial and epiphytic orchids. *Journal of Pollination Ecology* 32: 174–185.

Renner, S.S. (2006) Rewardless flowers in the angiosperms and the role of insect cognition in their evolution. In Waser, N.M. and Ollerton, J. (eds), *Plant–Pollinator Interactions: From Specialization to Generalization*. University of Chicago Press, Chicago, IL, pp. 123–144.

Rodríguez-Robles, J.A., Meléndez, E.J. and Ackerman, J.D. (1992) Effects of display size, flowering phenology, and nectar availability on effective visitation frequency in *Comparettia falcata* (Orchidaceae). *American Journal of Botany* 79: 1009–1017.

Roulston T.H. and Cane, J.H. (2000) Pollen nutritional content and digestibility for animals. *Plant Systematics and Evolution* 222: 187–209.

Sazima, M., Vogel, S., Prado, A.L., De Oliveira, D.M., Franz, G. and Sazima, I. (2001) The sweet jelly of *Combretum lanceolatum* flowers (Combretaceae): a cornucopia resource for bird pollinators in the Pantanal, western Brazil. *Plant Systematics and Evolution* 227: 195–208.

Schondube, J.E. and del Rio, C.M. (2003) Concentration-dependent sugar preferences in nectar-feeding birds: mechanisms and consequences. *Functional Ecology* 17: 445–453.

Sérsic, N.A. and Cocucci, A.A. (1996) A remarkable case of ornithophily in *Calceolaria*: food bodies as rewards for a non-nectarivorous bird. *Botanica Acta* 109: 172–176.

Salguero-Faria, J.A and Ackerman, J.D. (1999) A nectar reward: is more better? *Biotropica* 31: 303–311.

Singer, R.B. and Sazima, M. (2000) The pollination of *Stenorrhynchos lanceolatus* (Aublet) L. C. Rich. (Orchidaceae: Spirathinae) by hummingbirds in southeastern Brazil. *Plant Systematics and Evolution* 223: 221–227.

Waser, N.M. and Price, M.V. (1993) Crossing distance effects on prezygotic performance in plants: an argument for female choice. *Oikos* 68: 303–308.

Waser, N.M., CaraDonna, P.J. and Price, M.V. (2018) Atypical flowers can be as profitable as typical hummingbird flowers. *The American Naturalist* 192: 644–653.

Wester, P. and Classen-Bockhoff, R. (2007) Floral diversity and pollen transfer mechanisms in bird-pollinated *Salvia* species. *Annals of Botany* 100: 401–421.

Wooller, R.D., Richardson, K.C. and Pagendham, C.M. (1988) The digestion of pollen by some Australian birds. *Australian Journal of Zoology* 36: 357–362.

Young, F.J. and Montgomery, S.H. (2020) Pollen feeding in *Heliconius* butterflies: the singular evolution of an adaptive suite. *Proceedings of the Royal Society B* 287: 20201304. https://doi.org/10.1098/rspb.2020.1304

Chapter 7 – Misaligned interests

Ali, S.A. (1931) The role of sunbirds and flower-peckers in the propagation and distribution of the tree-parasite, *Loranthus longiflorus* Dest., in the Konkan (W. India). *Journal of the Bombay Natural History Society* 35: 144–149.

Barlow, B.A. and Wiens, D. (1977) Host-parasite resemblance in Australian mistletoes: the case for cryptic mimicry. *Evolution* 31: 69–84.

Bernhardt, P. and Calder, D.M (1981) The floral ecology of sympatric populations of *Amyema pendulum* and *Amyema quandang* (Loranthaceae). *Bulletin of the Torrey Botanical Club* 108: 213–230.

Blick, R.A.J., Burns, K.C. and Moles, A.T. (2012) Predicting network topology of mistletoe–host interactions: do mistletoes really mimic their hosts? *Oikos* 121(5), 761–771.

Boaventura, M.G., Villamil, N., Teixido, A.L. *et al.* (2022) Revisiting florivory: an integrative review and global patterns of a neglected interaction. *New Phytologist* 233: 132–144.

Canyon, D.V. and Hill, C.J. (1997) Mistletoe host-resemblance: a study of herbivory, nitrogen and moisture in two Australian mistletoes and their host trees. *Australian Journal of Ecology* 22: 395–403.

Colwell, R.K, Rangel, T.F., Fučiková, K., Sustaita, D., Yanega, G.M. and Rico-Guevara, A. (2023) Repeated evolution of unorthodox feeding styles drives a negative correlation between foot size and bill length in hummingbirds. *The American Naturalist*. https://doi.org/10.1086/726036. See also: www.science.org/content/article/some-hummingbirds-are-flower-robbers-here-s-how-spot-them

Cotton, P.A. (1998) Temporal partitioning of a floral resource by territorial hummingbirds. *Ibis* 140: 647–653.

Crates, R., Watson, D.M., Albery G.F. *et al.* (2022) Mistletoes could moderate drought impacts on birds, but are themselves susceptible to drought-induced dieback. *Proceedings of the Royal Society B* 289: 20220358. https://doi.org/10.1098/rspb.2022.0358

Davidar, P. (1983) Similarity between flowers and fruits in some flowerpecker pollinated mistletoes. *Biotropica* 15: 32–37.

Ladley, J.L., Kelly, D. and Robertson, A.W. (1997) Explosive flowering, nectar production, breeding systems, and pollinators of New Zealand mistletoes (Loranthaceae). *New Zealand Journal of Botany* 35: 345–360.

Maloof, J.E. and Inouye, D.W. (2000) Are nectar robbers cheaters or mutualists? *Ecology* 81: 2651–2661.

Nickrent, D. (2020) Parasitic angiosperms: how often and how many? *Taxon* 69: 5–27.

Oborny, B. (1991) Criticisms on optimal foraging in plants: a review. *Abstracta Botanica* 15: 67–76.

Ollerton, J. and Lack, A.J. (1996) Partial predispersal seed predation in *Lotus corniculatus* L. (Fabaceae). *Seed Science Research* 6: 65–69.

Ollerton, J. and Nuttman, C. (2013) Aggressive displacement of *Xylocopa nigrita* carpenter bees from flowers of *Lagenaria sphaerica* (Cucurbitaceae) by territorial male Eastern Olive Sunbirds (*Cyanomitra olivacea*) in Tanzania. *Journal of Pollination Ecology* 11: 21–26.

Ollerton, J., Rouquette, J.R. and Breeze, T.D. (2016) Insect pollinators boost the market price of culturally important crops: holly, mistletoe and the spirit of Christmas. *Journal of Pollination Ecology* 19: 93–97.

Pyke, G.H. (1984) Optimal foraging theory: a critical review. *Annual Review of Ecology and Systematics* 15: 523–575.

Pyke, G.H. (2016) Plant–pollinator co-evolution: it's time to reconnect with optimal foraging theory and evolutionarily stable strategies. *Perspectives in Plant Ecology, Evolution and Systematics* 19: 70–76.

Samu, F. (1991) Optimal plant foraging and the marginal value theorem: a zoologist's point of view. *Abstracta Botanica* 15: 77–81.

Singh, V.K., Barman, C. and Tandon, R. (2014) Nectar robbing positively influences the reproductive success of *Tecomella undulata* (Bignoniaceae). *PLoS One* 9: e102607. https://doi.org/10.1371/journal.pone.0102607

Watson D.M. (2001) Mistletoe: a keystone resource in forests and woodlands worldwide. *Annual Review of Ecology and Systematics* 32: 219–249.

Watson, D.M., McLellan, R.C. and Fontúrbel, F.E. (2022) Functional roles of parasitic plants in a warming world. *Annual Review of Ecology, Evolution, and Systematics* 53: 25–45.

Watts, S., Huamán Ovalle, D., Moreno Herrera, M. and Ollerton, J. (2012) Pollinator effectiveness of native and non-native flower visitors to an apparently generalist Andean shrub, *Duranta mandonii* (Verbenaceae). *Plant Species Biology* 27: 147–158.

Chapter 8 – Senses and sensitivities

Altshuler, D.L. and Wylie, D.R. (2020) Hummingbird vision. *Current Biology* 30: R103–R105.

Anon. (n.d.) Thousands of honeyeaters migrate. www.lfwseq.org.au/thousands-honeyeaters-migrate

Audubon, J.J. (1826) Account of the habits of the turkey buzzard, *Vultur aura*, particularly with the view of exploding the opinion generally entertained of its extraordinary power of smelling. *The Edinburgh New Philosophical Journal* 2: 172–184.

Bennett, A. and Théry, M. (2007) Avian color vision and coloration: multidisciplinary evolutionary biology. *The American Naturalist* 169: S1–S6.

Bent, A.C. (1953) *Life Histories of North American Wood Warblers*. Smithsonian Institution United States National Museum Bulletin 203 (reprinted by Dover Editions 1963). Available at Project Gutenberg: www.gutenberg.org/files/40100/40100-h/40100-h.htm

Bergamo, P., Rech, A.R., Brito, V.L.G. and Sazima, M. (2016) Flower colour and visitation rates of *Costus arabicus* support the 'bee avoidance' hypothesis for red-reflecting hummingbird-pollinated flowers. *Functional Ecology* 30: 710–720.

Brice, A.T. (1992) The essentiality of nectar and arthropods in the diet of the Anna's hummingbird (*Calypte anna*). *Comparative Biochemistry and Physiology Part A: Physiology* 101: 151–155.

Burd, M., Stayton, C.T., Shrestha, M. and Dyer, A.G. (2014) Distinctive convergence in Australian floral colours seen through the eyes of Australian birds. *Proceedings of the Royal Society B* 281: 220132862.

Chen, Z., Niu, Y., Liu, C.-Q. and Sun, H. (2020) Red flowers differ in shades between pollination systems and across continents. *Annals of Botany* 126: 837–848.

Chittka, L. and Niven, J. (2009) Are bigger brains better? *Current Biology* 19: R995–R1008.

Chittka, L. and Waser, N.M. (1997) Why red flowers are not invisible to bees. *Israel Journal of Plant Sciences* 45: 169–183.

Corfield, J.R., Price, K., Iwaniuk, A.N., Gutiérrez-Ibáñez, C., Birkhead, T. and Wylie, D.R. (2015) Diversity in olfactory bulb size in birds reflects allometry, ecology, and phylogeny. *Frontiers in Neuroanatomy* 9: 102. https://doi.org/10.3389/fnana.2015.00102

Cotton, P.A. (2007) Seasonal resource tracking by Amazonian hummingbirds. *Ibis* 149: 135–142.

Craig, A.J.F.K. and Hulley, P.E. (1994) Sunbird movements: a review, with possible models. *Ostrich* 65: 106–110.

Darwin, C. (1891) *Journal of Researches into the Natural History and Ecology of the Countries Visited During the Voyage of H.M.S. Beagle Round the World, Under the Command of Capt. Fitz Roy, R.N.* Ward, Lock and Co., London.

Goldsmith, K.M. and Goldsmith, T.H. (1982) Sense of smell in the black-chinned hummingbird. *The Condor* 84: 237–238.

Hart, N.S. (2001) The visual ecology of avian photoreceptors. *Progress in Retinal and Eye Research* 20: 675–703.

Heringer, H., Palmeira, L.R.M., Alves, A.C.F., Jacobi, C.M., Fontenelle, J.C.R. and Martins, R.P. (2005) Estudo da capacidade olfatória em três representantes da subfamília Trochilinae: *Eupetomena macroura* (Gould, 1853), *Thalurania furcata eriphile* (Lesson, 1832) e *Amazilia lacteal* (Lesson, 1832). *Proceedings VII Congress of the Brazilian Society of Ecologia.* www.seb-ecologia.org.br/revistas/indexar/anais/viiceb/resumos/318a.pdf

Heystek, A., Geerts, S., Barnard, P. and Pauw, A. (2014) Pink flower preference in sunbirds does not translate into plant fitness differences in a polymorphic *Erica* species. *Evolutionary Ecology* 28: 457–470. https://doi.org/10.1007/s10682-014-9693-z

Hilpman, E.T. and Busch, J.W. (2021) Floral traits differentiate pollination syndromes and species but fail to predict the identity of floral visitors to *Castilleja. American Journal of Botany* 108: 2150–2161.

Hu, H., Wang, Y., Fabbri, M. *et al.* (2023) Cranial osteology and palaeobiology of the Early Cretaceous bird *Jeholornis prima* (Aves: Jeholornithiformes). *Zoological Journal of the Linnean Society* 198: 93–112. https://doi.org/10.1093/zoolinnean/zlac089. See also: https://phys.org/news/2022-10-fossil-bird-skull-reconstruction-reveals.html

Knudsen, J.T., Tollsten, L., Groth, I., Bergström, G. and Raguso, R.A. (2004) Trends in floral scent chemistry in pollination syndromes: floral scent composition in hummingbird-pollinated taxa. *Botanical Journal of the Linnean Society* 146: 191–199.

Laursen K., Holm, E. and Sørensen, I. (1997) Pollen as a marker in migratory warblers, Sylviidae. *Ardea* 85: 223–231.

Levey, D.J. and Stiles, F.G. (1992) Evolutionary precursors of long-distance migration: resource availability and movement patterns in Neotropical landbirds. *The American Naturalist* 140: 447–476.

Low, T. (2016) *Where Song Began: Australia's Birds and How They Changed the World*. Yale University Press, New Haven, CT.

Lunau, K., Papiorek, S., Eltz, T. and Sazima, M. (2011) Avoidance of achromatic colours by bees provides a private niche for hummingbirds. *Journal of Experimental Biology* 214: 1607–1612.

Martin, G.R. (2020) *Bird Senses: How and What Birds See, Hear, Smell, Taste, and Feel*. Pelagic Publishing, Exeter.

Moore, R.T. (1947) Habits of male hummingbirds near their nests. *The Wilson Bulletin* 59: 21–25.

Moran, A.J., Prosser, S.W.J. and Moran, J.A. (2019) DNA metabarcoding allows non-invasive identification of arthropod prey provisioned to nestling rufous hummingbirds (*Selasphorus rufus*). *PeerJ* 7: e6596.

Núñez, P., Méndez, M. and López-Rull, I. (2021) Can foraging hummingbirds use smell? a test with the Amazilia Hummingbird *Amazila amazilia* [*sic*]. *Ardeola* 68: 433–444.

Ohashi, K. and Thomson, J.D. (2009) Trapline foraging by pollinators: its ontogeny, economics and possible consequences for plants. *Annals of Botany* 103: 1365–1378.

Pennisi, E. (2021) Textbooks say most birds can't smell. Scientists are proving them wrong. www.sciencemag.org/news/2021/07/textbooks-say-most-birds-cant-smell-scientists-are-proving-them-wrong

Porsch, O. (1924) Zukunftsaufgaben der Vogelblumenforschung auf Grund neuesten Tatbestandes. *Die Naturwissenschaften* 12: 993–1000.

Porsch, O. (1931) Grellrot als vogelblumenfarbe. *Biologia Generalis* 7: 647–674.

Poulin, B., Lefebvre, G. and McNeil, R. (1994) Diets of land birds from northeastern Venezuela. *The Condor* 96: 354–367.

Pyke, G.H. (1980), The foraging behaviour of Australian honeyeaters: a review and some comparisons with hummingbirds. *Australian Journal of Ecology* 5: 343–369.

Rathcke, B.J. (2000) Birds, pollination reliability, and green flowers in an endemic island shrub, *Pavonia bahamensis* (Malvaceae). *Rhodora* 102: 392–414.

Rehkämper, G., Schuchmann K.-L., Schleicher, A. and Zilles K. (1991) Encephalization in hummingbirds (Trochilidae). *Brain, Behavior and Evolution* 37: 85–91.

Remsen, J.V., Stiles, F.G. and Scott, P.E. (1986) Frequency of arthropods in stomachs of tropical hummingbirds. *The Auk* 103: 436–441.

Rodríguez-Gironés, M.A. and Santamaría, L. (2004) Why are so many bird flowers red? *PLoS Biology* 2: 1515–1519.

Ruschi, P.A. (2014) Frugivory by the hummingbird *Chlorostilbon notatus* (Apodiformes: Trochilidae) in the Brazilian Amazon. *Boletim do Museu de Biologia Mello Leitão (Nova Série)* 35: 43–47.

Shrestha, M., Dyer, A.G., Boyd-Gerny, S., Wong, B.B.M. and Burd, M. (2013) Shades of red: bird-pollinated flowers target the specific colour discrimination abilities of avian vision. *New Phytologist* 198: 301–310.

Symes, C.T., Nicolson, S.W. and McKechnie, A.E. (2008) Response of avian nectarivores to the flowering of *Aloe marlothii*: a nectar oasis during dry South African winters. *Journal of Ornithology* 149: 13–22.

Tello-Ramos, M.C., Hurly, T.A. and Healy, S.D. (2015) Traplining in hummingbirds: flying short-distance sequences among several locations. *Behavioral Ecology* 26: 812–819.

Toda, Y., Ko, M.-C., Liang, Q. *et al.* (2021) Early origin of sweet perception in the songbird radiation. *Science* 373: 226–231.

Torres-Vanegas, F., Hadley, A.S., Kormann, U.G., Jones, F.A., Betts, M.G. and Wagner, H.H. (2019) The landscape genetic signature of pollination by trapliners: evidence from the tropical herb, *Heliconia tortuosa*. *Frontiers in Genetics* 10: 1206. https://doi.org /10.3389/fgene.2019.01206

van Balen, B. (2020) Swinhoe's White-eye (*Zosterops simplex*), version 1.0. *Birds of the World*. Cornell Lab of Ornithology, Ithaca, NY. https://doi.org/10.2173/bow.swiwhe1.01

van der Kooi, C.J., Elzenga, J.T.M., Staal, M. and Stavenga, D.G. (2016) How to colour a flower: on the optical principles of flower coloration. *Proceedings of the Royal Society B* 28320160429. https://doi.org/10.1098/rspb.2016.0429

van der Kooi, C.J., Dyer, A.G., Kevan, P.G. and Lunau, K. (2019) Functional significance of the optical properties of flowers for visual signalling. *Annals of Botany* 123: 263–276.

Wagner, H.O. (1946) Food and feeding habits of Mexican hummingbirds. *The Wilson Bulletin* 58: 69–93.

Waser, N.M. (1979) Pollinator availability as a determinant of flowering time in ocotillo (*Fouquieria splendens*). *Oecologia* 39: 107–121.

Wendelken, P.W. and Martin, R.F. (1988) Avian consumption of the fruit of the cacti *Stenocereus eichlamii* and *Pilosocereus maxonii* in Guatemala. *The American Midland Naturalist* 119: 235–43.

Yanega, G. and Rubega, M. (2004) Hummingbird jaw bends to aid insect capture. *Nature* 428: 615. https://doi.org/10.1038/428615a

Young, A.M. (1971) Foraging for insects by a tropical hummingbird. *The Condor* 73: 36–45.

Chapter 9 – Codependent connections

Alarcón, R., Waser, N.M. and Ollerton, J. (2008) Year-to-year variation in the topology of a plant–pollinator interaction network. *Oikos* 117: 1796–1807.

Bascompte, J. and Jordano, P. (2014) *Mutualistic Networks*. Princeton University Press, Princeton, NJ.

Blak, K. (2021) *The Essential Companion to Talking Therapy*. Watkins Publishing, London.

Bustos, A., Wüest, R.O., Graham, C.H. and Varassin, I.G. (2023) The effect of species role and trait-matching on plant fitness in a plant–hummingbird interaction network. *Flora* 305: 152348.

CaraDonna, P.J., Petry, W.K., Brennan, R.M. *et al.* (2017) Interaction rewiring and the rapid turnover of plant–pollinator networks. *Ecology Letters* 20: 385–394.

Clements, R.E. and Long, F.L. (1923) *Experimental Pollination: an Outline of the Ecology of Flowers and Insects*. Carnegie Institute of Washington, Washington, DC.

Dalsgaard, B., Martín González, A.M., Olesen, J.M. Timmermann, A., Andersen, L.H. and Ollerton, J. (2008) Pollination networks and functional specialization: a test using Lesser Antillean plant–hummingbird assemblages. *Oikos* 117: 789–793.

Dalsgaard, B., Martín González, A.M., Olesen, J.M. *et al.* (2009) Plant–hummingbird interactions in the West Indies: floral specialisation gradients associated with environment and hummingbird size. *Oecologia* 159: 757–766.

Duchenne, F., Wüest, R.O. and Graham, C.H. (2022) Seasonal structure of interactions enhances multidimensional stability of mutualistic networks. *Proceedings of the Royal Society B* 289: 20220064.

Gavini, S.S., Sáez, A., Tur, C. and Aizen, M. (2021) Pollination success increases with plant diversity in high-Andean communities. *Scientific Reports* 11: 22107. doi.org/10.1038/s41598-021-01611-w

Graham, C.H., Parra, J.L., Rahbek, C. and McGuire, J.A. (2009) Phylogenetic structure in tropical hummingbird communities. *Proceedings of the National Academy of Sciences of the USA* 106: 19673–19678. doi.org/10.1073/pnas.0901649106

Interaction Web Database (2023) www.ecologia.ib.usp.br/iwdb.

Janeček, Š., Chmel, K., Mlíkovský, J. *et al.* (2022) Spatiotemporal pattern of specialization of sunbird–plant networks on Mt. Cameroon. *Oecologia* 199: 885–896.

Jønsson, K.A., Thomassen, E.E., Iova, B., Sam, K. and Thomsen, P.F. (2023) Using environmental DNA to investigate avian interactions with flowering plants. *Environmental DNA* 5: 462– 475.

Leimberger, K.G., Hadley, A.S. and Betts, M.G. (2023) Plant–hummingbird pollination networks exhibit limited rewiring after experimental removal of a locally abundant plant species. *Journal of Animal Ecology* 00, 1–15. https://doi.org/10.1111/1365-2656.13935

Lopes, S.A., Bergamo, P.J., Queiroz, S.N.P., Ollerton, J., Santos, T. and Rech, A.R. (2022) Heterospecific pollen deposition is positively associated with reproductive success in a diverse hummingbird-pollinated plant community. *Oikos* e08714. https://doi.org/10.4444/oik.08714

Mitchell, R.J., Flanagan, R.J., Brown, B.J. Waser, N.M. and Karron, J.D. (2009) New frontiers in competition for pollination. *Annals of Botany* 103: 1403–1413.

Møller-Stranges, F., Hedegaard, L.M. and Dalsgaard, B. (2021) Double mutualism between the Black-crowned Palm-Tanager (Passeriformes, Pheanicophilidae, *Phaenicophilus palmarum*) and the Beach Creeper (Rubiaceae, *Ernodea littoralis*) on Hispaniola, Greater Antilles, Caribbean. *Caribbean Journal of Science* 51: 86–91.

Nsor, C.A., Godsoe, W. and Chapman, H.M. (2019) Promiscuous pollinators: evidence from an Afromontane sunbird–plant pollen transport network. *Biotropica* 51: 538–548.

Ollerton, J., Johnson, S.D., Cranmer, L. and Kellie, S. (2003) The pollination ecology of an assemblage of grassland asclepiads in South Africa. *Annals of Botany* 92: 807–834.

Vitória, R.S., Vizentin-Bugoni, J. and Duarte, L.D.S. (2018) Evolutionary history as a driver of ecological networks: a case study of plant–hummingbird interactions. *Oikos* 127: 561–569.

Vizentin-Bugoni, J., Maruyama, P.K. and Sazima M. (2014) Processes entangling interactions in communities: forbidden links are more important than abundance in a hummingbird–plant network. *Proceedings of the Royal Society B* 281: 20132397.

Vizentin-Bugoni, J., Maruyama, P.K., Debastiani, V.J., Duarte, L.D.S., Dalsgaard, B. and Sazima, M. (2016) Influences of sampling effort on detected patterns and structuring processes of a Neotropical plant–hummingbird network. *Journal of Animal Ecology* 85: 262–272.

Vizentin-Bugoni J., Maruyama, P.K.M., Souza, C.S., Ollerton, J., Rech, A.R. and Sazima, M. (2018) Plant–pollinator networks in the tropics: a review. In Dáttilo, W. and Rico-Gray, V. (eds), *Ecological Networks in the Tropics.* Springer, Dordrecht, pp. 73–91.

Watts, S., Dormann, C.F., Martín González, A.M. and Ollerton, J. (2016) The influence of floral traits on specialisation and modularity of plant–pollinator

networks in a biodiversity hotspot in the Peruvian Andes. *Annals of Botany* 118: 415–429.
Zanata, T.B., Dalsgaard, B., Passos, F.C. *et al.* (2017) Global patterns of interaction specialization in bird–flower networks. *Journal of Biogeography* 44: 1891–1910.

Chapter 10 – Hitchhikers, drunks and killers

Bates, H.W. (1863) *The Naturalist on the River Amazons*. John Murray, London.
Belisle, M., Peay, K.G. and Fukami, T. (2012) Flowers as islands: spatial distribution of nectar-inhabiting microfungi among plants of *Mimulus aurantiacus*, a hummingbird-pollinated shrub. *Microbial Ecology* 63: 711–718. https://doi.org/10.1007/s00248-011-9975-8
Bizzarri, L., Baer, C.S. and Garcia-Robledo, C. (2022) DNA barcoding reveals generalization and host overlap in hummingbird flower mites: implications for the Mating Rendezvous Hypothesis. *The American Naturalist* 199: 576–583. https://doi.org/10.1086/718474. See also: https://ibol.org/barcodebulletin/research/2022-11-02-specialists-to-generalists-in-tropical-ecosystems/
Brooks, D.M. (2012) Birds caught in spider webs: a synthesis of patterns. *The Wilson Journal of Ornithology* 124: 345–353. https://doi.org/10.1676/11-148.1
Carignan, J.M. (1988) Predation on Rufous Hummingbird by praying mantid. *Texas Journal of Science* 40: 111.
Colwell, R.K. (1986) Community biology and sexual selection: lessons from hummingbird flower mites. In Diamond, J. and Case, T.J. (eds), *Community Ecology*. Harper and Row, New York, pp. 406–424.
Colwell, R.K. (1995) Effects of nectar consumption by the hummingbird flower mite *Proctolaelaps kirmsei* on nectar availability in *Hamelia patens*. *Biotropica* 27: 206–217.
Da Cruz, D.D., De Abreu, V.H.R. and Van Sluys, M. (2007) The effect of hummingbird flower mites on nectar availability of two sympatric *Heliconia* species in a Brazilian Atlantic Forest. *Annals of Botany* 100: 581–588.
da Silva, L.P., Pereira Coutinho, A., Heleno, R.H., Tenreiro, P.Q. and Ramos, J.A. (2016) Dispersal of fungi spores by non-specialized flower-visiting birds. *Journal of Avian Biology* 47: 438–442. https://doi.org/10.1111/jav.00806
de Alcantara Viana, J.V., Massufaro Giffu, M. and Hachuy-Filho, L. (2021) The silence of prey: hummingbirds do not respond to potential ambush predators on flowers. *Austral Ecology* 46: 515–520.
de Carvalho, W.D., Norris, D. and Michalski, F. (2016) Opportunistic predation of a Common Scale-backed Antbird (*Willisornis poecilinotus*) by a Goliath bird-eating spider (*Theraphosa blondi*) in the Eastern Brazilian Amazon. *Studies on Neotropical Fauna and Environment* 51: 239–241. https://doi.org/10.1080/01650521.2016.1237802
Grant, J. 1(959) Hummingbirds attacked by wasps. *Canadian Field-Naturalist* 73: 174.
Herrera, C.M., de Vega, C., Canto, A. and Pozo, M.J. (2009) Yeasts in floral nectar: a quantitative survey. *Annals of Botany* 103: 1415–1423.
Hofslund, P.B. (1977) Dragonfly attacks and kills a Ruby-throated Hummingbird. *Loon* 49: 238.
Joharchi, O., Vorontsov, D.D. and Walter, D.E. (2021) Oldest determined record of a mesostigmatic mite (Acari: Mesostigmata: Sejidae) in Cretaceous Burmese amber. *Acarologia* 61: 641–649. https://doi.org/10.24349/goj5-BZms

Lara, C. and Ornelas, J.F. (2001) Nectar 'theft' by hummingbird flower mites and its consequences for seed set in *Moussonia deppeana*. *Functional Ecology* 15: 78–84.

Lara, C. and Ornelas, J.F. (2003) Hummingbirds as vectors of fungal spores in *Moussonia deppeana* (Gesneriaceae): taking advantage of a mutualism? *American Journal of Botany* 90: 262–269.

Merian, M.S. (1705) *Metamorphosis Insectorum Surinamensium*. Gerard Valck, Amsterdam.

Mittelbach, M., Yurkov, A.M., Nocentini, D., Nepi, M., Weigend, M. and Begerow, D. (2015) Nectar sugars and bird visitation define a floral niche for basidiomycetous yeast on the Canary Islands. *BMC Ecology* 15: 2. https://doi.org/10.1186/s12898-015-0036-x

Noske, R.A. (1993) *Bruguiera hainesii*: another bird-pollinated mangrove? *Biotropica* 25: 481–483.

Nyffeler, M., Maxwell, M.R and Remsen, J.V. (2017) Bird predation by praying mantises: a global perspective. *The Wilson Journal of Ornithology* 129: 331–344.

O'Hanlon, R. (2016) Travelling naturalists. In van Delft, M. and Mulder, H. (eds), *Maria Sibylla Merian (1705) Metamorphosis Insectorum Surinamensium*. Facsimile edition by Lannoo Publishers and Koninklijke Bibliotheek, National Library of the Netherlands, in collaboration with Amsterdam University, pp. 7–8.

Proctor, H. and Owens, I. (2000) Mites and birds: diversity, parasitism and coevolution. *Trends in Ecology and Evolution* 15: 358–364.

Raguso, R.A. (2023) Hidden worlds within flowers. *Current Biology* 33: R506–R512. https://doi.org/10.1016/j.cub.2023.04.054

Theron-De Bruin, N., Dreyer, L.L., Ueckermann, E.A., Wingfield, M.J. and Roets, F. (2018) Birds mediate a fungus-mite mutualism. *Microbial Ecology* 75: 863–874. https://doi.org/10.1007/s00248-017-1093-9

University of Connecticut (2023) Biodiversity Research Collections. https://biodiversity.uconn.edu/rob-colwells-mites.

Vannette, R.L. and Fukami, T. (2018) Contrasting effects of yeasts and bacteria on floral nectar traits. *Annals of Botany* 121: 1343–1349.

Walther, B.A. (2016) Birds caught in spider webs in Asia. *Avian Research* 7: 16. https://doi.org/10.1186/s40657-016-0051-4

Whitehead, M. (2019) Wild yeasts are everywhere. Some of them will even make beer for you. https://michaelwhitehead.net/2019/05/14/wild-yeasts-are-everywhere-some-of-them-will-even-make-beer-for-you

Whitehead, M. (2020) Suburban Melbourne yeast capture. The rancid smell of failure. https://michaelwhitehead.net/2020/07/31/suburban-melbourne-yeast-capture-the-rancid-smell-of-failure

Chapter 11 – The limits to specialisation

Abrahamczyk, S. (2019) Comparison of the ecology and evolution of plants with a generalist bird pollination system between continents and islands worldwide. *Biological Reviews* 94: 1658–1671.

Abrahamczyk, S., Souto-Vilarós, D. and Renner, S.S. (2014) Escape from extreme specialization: passionflowers, bats and the sword-billed hummingbird. *Proceedings of the Royal Society B* 281: 20140888. https://doi.org/10.1098/rspb.2014.0888

Amorim, F.W., Ballarin, C.S., Mariano, G. *et al.* (2020) Good heavens what animal can pollinate it? A fungus-like holoparasitic plant potentially pollinated by opossums. *Ecology* 101: e03001. https://doi.org/10.1002/ecy.3001

Amorim, M.D., Maruyama, P.K., Baronio, G.J., Azevedo, C.S. and Rech, A.R. (2022) Hummingbird contribution to plant reproduction in the rupestrian grasslands is not defined by pollination syndrome. *Oecologia* 199: 1–12.

Amorim, F.W., Ballarin, C.S., Spicacci, G. *et al.* (2023) Opossums and birds facilitate the unexpected bat visitation to the ground-flowering *Scybalium fungiforme*. *Ecology* 104: e3935. https://doi.org/10.1002/ecy.3935

Amundsen, T. (2000) Why are female birds ornamented? *Trends in Ecology and Evolution* 15: 149–155.

Amundsen, T., Forsgren, E. and Hansen, L.T.T. (1997) On the function of female ornaments: male bluethroats prefer colourful females. *Proceedings of the Royal Society B* 264: 1579–1586.

Arteaga-Chávez, W.A., León-Reyes, A.E., Rodríguez, L., Hipo, D.P. and Alcívar, J.L.C. (2022) Primera descripción del nido y polluelos del Colibrí Picoespada *Ensifera ensifera* (Trochilidae). *Cotinga* 44: 126–132.

Barrett, S.C.H. (2013) The evolution of plant reproductive systems: how often are transitions irreversible? *Proceedings of the Royal Society B* 280: 20130913.

Beltrán, D.F., Araya-Salas, M., Parra, J.L. Stiles, F.G. and Rico-Guevara, A. (2022) The evolution of sexually dimorphic traits in ecological gradients: an interplay between natural and sexual selection in hummingbirds. *Proceedings of the Royal Society B* 289: 20221783. https://doi.org/10.1098/rspb.2022.1783

Castellanos, M.C., Wilson, P. and Thomson, J.D. (2004) 'Anti-bee' and 'pro-bird' changes during the evolution of hummingbird pollination in *Penstemon* flowers. *Journal of Evolutionary Biology* 17: 876–885.

Chmel, K., Ewome, F.L., Gómez, G.U. *et al.* (2021) Bird pollination syndrome is the plant's adaptation to ornithophily, but nectarivorous birds are not so selective. *Oikos* 130: 1411–1424.

Cotton, P.A. (1998) Coevolution in an Amazonian hummingbird–plant community. *Ibis* 140: 639–646.

Dalsgaard, B., Timmermann, A., Martín González, A.M., Olesen, J.M, Ollerton, J. and Andersen, L.H. (2012) *Heliconia*–hummingbird interactions in the Lesser Antilles: a geographic mosaic? *Caribbean Journal of Science* 46: 328–331.

Darwin, C. (1871) *The Descent of Man and Selection in Relation to Sex*. John Murray, London.

de Camargo, M.G.G., Lunau, K. Batalha, M.A. Brings, S. de Brito, V.L.G. and Morellato, L.P.C. (2019) How flower colour signals allure bees and hummingbirds: a community-level test of the bee avoidance hypothesis. *New Phytologist* 222: 1112–1122.

Diller C., Castañeda-Zárate M. and Johnson S.D. (2019) Generalist birds outperform specialist sunbirds as pollinators of an African *Aloe*. *Biology Letters* 15: 20190349.

Geerts, S. and Pauw, A. (2009) Hyper-specialization for long-billed bird pollination in a guild of South African plants: the Malachite Sunbird pollination syndrome. *South African Journal of Botany* 75: 699–706.

Hargreaves, A.L., Johnson, S.D. and Nol, E. (2004) Do floral syndromes predict specialization in plant pollination systems? An experimental test in an 'ornithophilous' African *Protea* . *Oecologia* 140: 295–301.

Janeček, Š., Chmel, K., Gómez, G.U. *et al.* (2020) Ecological fitting is a sufficient driver of tight interactions between sunbirds and ornithophilous plants. *Ecology and Evolution* 10: 1784– 1793.

Johnson, S.D. and Nicolson, S.W. (2008) Evolutionary associations between nectar properties and specificity in bird pollination systems. *Biology Letters* 4: 49–52. https://doi.org/10.1098/rsbl.2007.0496

Johnson, S.D. and Steiner, K.E. (2000) Generalization versus specialization in plant pollination systems. *Trends in Ecology and Evolution* 15: 140–143. https://doi.org/10.1016/S0169-5347(99)01811-X

Jones, I.L. and Hunter, F.M. (1993) Mutual sexual selection in a monogamous seabird. *Nature* 362: 238–239.

Jones, I.L. and Hunter, F.M. (1993) Experimental evidence for mutual inter- and intrasexual selection favouring a crested auklet ornament. *Animal Behaviour* 57: 521–528.

Maruyama P.K., Oliveira, G.M., Ferreira, C., Dalsgaard, B. and Oliveira, P.E. (2013) Pollination syndromes ignored: importance of non-ornithophilous flowers to Neotropical savanna hummingbirds. *Naturwissenschaften* 100: 1061–1068.

Mascó, M., Noy-Meir, I. and Sérsic, A.N. (2004) Geographic variation in flower color patterns within *Calceolaria uniflora* Lam. in Southern Patagonia. *Plant Systematics and Evolution* 244: 77–91.

McWhorter, T.J., Rader, J.A., Schondube, J.E. *et al.* (2021) Sucrose digestion capacity in birds shows convergent coevolution with nectar composition across continents. *iScience* 24: 102717. https://doi.org/10.1016/j.isci.2021.102717

Moore, R.T. (1947) Habits of male hummingbirds near their nests. *The Wilson Bulletin* 59: 21–25.

Muchhala, N. (2007) Adaptive trade-off in floral morphology mediates specialization for flowers pollinated by bats and hummingbirds. *The American Naturalist* 169: 494–504.

Ollerton, J., Killick, A., Lamborn, E., Watts, S. and Whiston, M. (2007) Multiple meanings and modes: on the many ways to be a generalist flower. *Taxon* 56: 717–728.

Paiaro, V., Cocucci, A.A., Oliva, G.E. and Sérsic, A.N. (2017) The role of facultatively nectarivorous birds as pollinators of *Anarthrophyllum desideratum* in the Patagonian steppe: a geographical approach. *Ecología Austral* 27: 312–325.

Remsen, J.V., Stiles, F.G. and Scott, P.E. (1986) Frequency of arthropods in stomachs of tropical hummingbirds. *The Auk* 103: 436–441.

Rico-Guevara, A. and Araya-Salas, M. (2015) Bills as daggers? A test for sexually dimorphic weapons in a lekking hummingbird. *Behavioral Ecology* 26: 21–29.

Rico-Guevara, A., Fan T.-H. and Rubega M.A. (2015) Hummingbird tongues are elastic micropumps. *Proceedings of the Royal Society B* 282: 20151014. See also the amazing videos of hummingbirds feeding on Alejandro Rico-Guevara's website: https://alejorico.com

Serrano-Serrano, M.L., Rolland, J., Clark J.L., Salamin, N. and Perret, M. (2017) Hummingbird pollination and the diversification of angiosperms: an old and successful association in Gesneriaceae. *Proceedings of the Royal Society B* 284: 20162816.

Sérsic, A.N., Mascó, M. and Noy-Meir, I. (2001) Natural hybridization between species of *Calceolaria* with different pollination syndromes in southern Patagonia, Argentina. *Plant Systematics and Evolution* 230: 111–124.

Sonne, J., Martín González, A.M., Maruyama, P.K. *et al.* (2016) High proportion of smaller ranged hummingbird species coincides with ecological specialization across the Americas. *Proceedings of the Royal Society B* 283: 20152512. https://doi.org/10.1098/rspb.2015.2512

Sonne, J., Vizentin-Bugoni, J., Maruyama, P.K. *et al.* (2020) Ecological mechanisms explaining interactions within plant–hummingbird networks: morphological matching increases towards lower latitudes. *Proceedings of the Royal Society B* 287: 20192873. https://doi.org/10.1098/rspb.2019.2873
Stiles, F.G. (1975) Ecology, flowering phenology and hummingbird pollination of some Costa Rican *Heliconia* species. *Ecology* 56: 285–301.
Stiles, F.G. (1982) Aggressive and courtship displays of the male Anna's hummingbird. *The Condor* 84: 208–225.
Stiles, F.G. and Wolf, L.L. (1979) Ecology and evolution of lek mating behaviour in the Long-tailed Hermit hummingbird. *Ornithological Monographs* 27: 1–77.
Temeles, E.J. and Bishop, G.A. (2019) A hurricane alters pollinator relationships and natural selection on an introduced island plant. *Biotropica* 51: 129–138.
Temeles, E.J. and Kress, J. (2003) Adaptation in a plant hummingbird association. *Science* 300: 630–633.
Temeles, E.J., Pan, I.L., Brennan, J.L. and Horwitt, J.N. (2000) Evidence for ecological causation of sexual dimorphism in a hummingbird. *Science* 289: 441–443.
Thompson, J.N. (2005) *The Geographic Mosaic of Coevolution.* University of Chicago Press, Chicago, IL.
Tripp, E.A. and Manos, P.S. (2008) Is floral specialization an evolutionary dead-end? Pollination system transitions in *Ruellia* (Acanthaceae). *Evolution* 62: 1712–1737.
Waser, N.M. and Price, M.V. (1983) Pollinator behaviour and natural selection for flower colour in *Delphinium nelsonii. Nature* 302: 422–424.
Waser, N.M. and Price, M.V. (1990) Pollination efficiency and effectiveness of bumblebees and hummingbirds visiting *Delphinium nelsonii. Collecteana Botanica* 19: 9–20.

Chapter 12 – Islands in the sea, islands in the sky

Alayón, D.I.O. (2013) The Macaronesian bird-flowered element as a model system to study the evolution of ornithophilous floral traits. *Vieraea* 41: 73–89.
Ashman, T.-L. and Schoen, D.J. (1994) How long should flowers live? *Nature* 371: 788–791.
Ashman, T.-L. and Schoen, D.J. (1996) Floral longevity: fitness consequences and resource costs. In Lloyd, D.G. and Barrett, S.C.H. (eds), *Floral Biology*. Springer, Boston, MA, pp. 112–139.
Basnett, S. and Ganesan, R. (2022) A comprehensive review on the taxonomy, ecology, reproductive biology, economic importance and conservation status of Indian Himalayan rhododendrons. *Botanical Review* 88: 505–544. https://doi.org/10.1007/s12229-021-09273-z
Basnett, S., Ganesan, R. and Devy, S.M. (2019) Floral traits determine pollinator visitation in *Rhododendron* species across an elevation gradient in the Sikkim Himalaya. *Alpine Botany* 129: 81–94.
Brown, E.D. and Hopkins, M.J.G. (1995) A test of pollinator specificity and morphological convergence between nectarivorous birds and rainforest tree flowers in New Guinea. *Oecologia* 103: 89–100.
Brown, E.D. and Hopkins, M.J.G. (1996) How New Guinea rainforest flower resources vary in time and space: implications for nectarivorous birds. *Australian Journal of Ecology* 21: 363–378.
Burbano, D.V., Valdivieso, J.C., Izurieta, J.C., Meredith, T.C. and Ferri, D.Q. (2022) 'Rethink and reset' tourism in the Galapagos Islands: stakeholders' views

on the sustainability of tourism development. *Annals of Tourism Research Empirical Insights* 3: 100057. https://doi.org/10.1016/j.annale.2022.100057

Cruden, R.W. (1972) Pollinators in high-elevation ecosystems: relative effectiveness of birds and bees. *Science* 176: 1439–1440.

Dalsgaard, B., Baquero, A.C., Rahbek, C., Olesen, J.M. and Wiley, J.W. (2016) Speciose opportunistic nectar-feeding avifauna in Cuba and its association to hummingbird island biogeography. *Journal of Ornithology* 157: 627–634.

de-Oliveira-Nogueira, C.H., Souza, U.F., Machado, T.M. *et al.* (2023) Between fruits, flowers and nectar: the extraordinary diet of the frog *Xenohyla truncata*. *Food Webs* 35: e00281. https://doi.org/10.1016/j.fooweb.2023.e00281

Dupont, Y.L. and Olesen, J.M. (2004) Fugleblomster på de Kanariske Øer. *Naturens Verden* 87: 2–11.

Dupont, Y.L. and Skov, C (2004) Influence of geographical distribution and floral traits on species richness of bees (Hymenoptera: Apoidea) visiting *Echium* species (Boraginaceae) of the Canary Islands. *International Journal of Plant Science* 165: 377–386.

Dupont, Y.L., Hansen, D.M. and Olesen, J.M. (2003) Structure of a plant flower–visitor network in the high-altitude sub-alpine desert of Tenerife, Canary Islands. *Ecography* 26: 301–310.

Dupont, Y.L., Hansen, D.M., Rasmussen, J.T. and Olesen, J.M. (2004) Evolutionary changes in nectar sugar composition associated with switches between bird and insect pollination: the Canarian bird-flower element revisited. *Functional Ecology* 18: 670–676. https://doi.org/10.1111/j.0269-8463.2004.00891.x

Fernández de Castro, A.G., Moreno-Saiz, J.C. and Fuertes-Aguilar, J. (2017) Ornithophily for the nonspecialist: differential pollination efficiency of the Macaronesian island paleoendemic *Navaea phoenicea* (Malvaceae) by generalist passerines. *American Journal of Botany* 104: 1556–1568.

Grant, B.R. and Grant, P.R. (1981) Exploitation of *Opuntia* cactus by birds on the Galápagos. *Oecologia* 49: 179–187.

Grant, P.R. (1993) Hybridization of Darwin's finches on Isla Daphne Major, Galapagos. *Philosophical Transactions of the Royal Society: Biological Sciences* 340: 127–139.

Grant, P.R and Grant, B.R (2010) Ecological insights into the causes of an adaptive radiation from long-term field studies of Darwin's finches. In Billick, I. and Price, M.V. (eds), *The Ecology of Place: Contributions of Place-Based Research to Ecological Understanding*. University of Chicago Press, Chicago, IL, pp. 109–133.

Higgins, P.J., Christidis, L. and Ford, H. (2020) Long-bearded Melidectes (*Melidectes princeps*), version 1.0. *Birds of the World*. Cornell Lab of Ornithology, Ithaca, NY. https://doi.org/10.2173/bow.lobmel1.01

Huang, Z.H., Song, Y.P. and Huang, S.Q. (2017) Evidence for passerine bird pollination in *Rhododendron* species. *AoB Plants* 9: 1–10.

Lack, D. (1947) *Darwin's Finches*. Cambridge University Press, Cambridge.

Lack, D. (1973) The numbers of species of hummingbirds in the West Indies. *Evolution* 27: 326–337.

Lamichhaney, S., Han, F., Webster, M.T., Andersson, L., Grant, B.R. and Grant, P.R. (2018) Rapid hybrid speciation in Darwin's finches. *Science* 359: 224–228. https://doi.org/10.1126/science.aao459

Milne, R.I. and Abbott, R.J. (2002) The origin and evolution of Tertiary relict floras. *Advances in Botanical Research* 38: 281–314.

Ojeda, D.I. (201). The Macaronesian bird-flowered element as a model system to study the evolution of ornithophilous floral traits. *Vieraea* 41: 73–89.

Ojeda, D.I., Santos-Guerra, A., Oliva-Tejera, F. *et al.* (2013) Bird-pollinated Macaronesian *Lotus* (Leguminosae) evolved within a group of entomophilous ancestors with post-anthesis flower color change. *Perspectives in Plant Ecology, Evolution and Systematics* 15: 193–204. https://doi.org/10.1016/j.ppees.2013.05.002

Ojeda, D.I., Valido, A. Fernández de Castro, A.G. *et al.* (2016) Pollinator shifts drive petal epidermal evolution on the Macaronesian Islands bird-flowered species. *Biology Letters* 12: 20160022.

Ojeda, I., Santos-Guerra, A., Jaén-Molina, R., Oliva-Tejera, F., Caujapé-Castells, J. and Cronk, Q. (2012) The origin of bird pollination in Macaronesian *Lotus* (Loteae, Leguminosae). *Molecular Phylogenetics and Evolution* 62: 306–318. https://doi.org/10.1016/j.ympev.2011.10.001

Olesen, J.M. (1985) The Macaronesian bird–flower element and its relation to bird and bee opportunists. *Botanical Journal of the Linnean Society* 91: 395–414.

Olesen, J.M. and Valido, A. (2003) Lizards as pollinators and seed dispersers: an island phenomenon. *Trends in Ecology and Evolution* 18: 177–181.

Olesen, J.M. and Valido, A. (2003) Bird pollination in Madeira Island. *Ardeola* 50: 67–69.

Olesen, J.M., Alarcón, M., Ehlers, B.K., Aldasoro, J.J. and Roquet, C. (2012) Pollination, biogeography and phylogeny of oceanic island bellflowers (Campanulaceae). *Perspectives in Plant Ecology, Evolution and Systematics* 14: 169–182.

Ollerton, J., Cranmer, L., Stelzer, R., Sullivan, S. and Chittka, L. (2009) Bird pollination of Canary Island endemic plants. *Naturwissenschaften* 96: 221–232.

Ollerton, J., Koju, N.P., Maharjan, S.R. and Bashyal, B. (2019) Interactions between birds and flowers of *Rhododendron* spp., and their implications for mountain communities in Nepal. *Plants, People, Planet* 2: 320–325. https://doi.org/10.1002/ppp3.10091

Ortega-Olivencia, A., Rodríguez-Riaño, T., Pérez-Bote, J.L. *et al.* (2012) Insects, birds and lizards as pollinators of the largest-flowered *Scrophularia* of Europe and Macaronesia. *Annals of Botany* 109: 153–167. https://doi.org/10.1093/aob/mcr255

Rodríguez-Rodríguez, M.C. and Valido, A. (2008) Opportunistic nectar-feeding birds are effective pollinators of bird-flowers from Canary Islands: experimental evidence from *Isoplexis canariensis* (Scrophulariaceae). *American Journal of Botany* 95: 1408–1415.

Rodríguez-Rodríguez, M.C. and Valido, A. (2011) Consequences of plant–pollinator and floral–herbivore interactions on the reproductive success of the Canary Islands endemic *Canarina canariensis* (Campanulaceae). *American Journal of Botany* 98: 1465–1474.

Sangster, G., Rodríguez-Godoy, F., Roselaar, C.S., Robb, M.S. and Luksenburg, J.A. (2016) Integrative taxonomy reveals Europe's rarest songbird species, the Gran Canaria blue chaffinch *Fringilla polatzeki*. *Journal of Avian Biology* 47: 159–166.

Shankar, A., Schroeder, R.J., Wethington, S.M., Graham, C.H. and Powers, D.R. (2020) Hummingbird torpor in context: duration, more than temperature, is the key to nighttime energy savings. *Journal of Avian Biology* 51: e02305. https://doi.org/10.1111/jav.02305

Snyder, N.F.R., Raffaele, H., Kirkconnell, A. and Wunderle, J. (2019) James W. Wiley, 1943–2018. *The Auk* 136: ukz037. https://doi.org/10.1093/auk/ukz037

Stelzer, R., Ollerton, J. and Chittka, L. (2007) Kein Nachweis für Hummelbesuch der Kanarischen Vogelblumen (Hymenoptera: Apidae). *Entomologia Generalis* 30: 153–154. https://doi.org/10.1127/entom.gen/30/2007/153

Tøttrup, A., Larsen, J. and Burgess, N. (2004) A first estimate of the population size of Loveridge's Sunbird *Nectarinia loveridgei*, endemic to the Uluguru Mountains, Tanzania. *Bird Conservation International* 14: 25–32.

Traveset, A., Olesen, J., Nogales, M. *et al.* (2015) Bird–flower visitation networks in the Galápagos unveil a widespread interaction release. *Nature Communications* 6, 6376. https://doi.org/10.1038/ncomms7376

Valido, A., Dupont, Y.L. and Hansen, D.M. (2002) Native birds and insects, and introduced honeybees visiting *Echium wildpretii* (Boraginaceae) in the Canary Islands. *Acta Oecologica* 23: 413–419.

Valido, A., Dupont, Y.L. and Olesen, J.M. (2004) Bird–flower interactions in the Macaronesian islands. *Journal of Biogeography* 31: 1945–1953.

Vogel, S. (1954) Blütenbiologische Typen als Elemente der Sippengliederung. *Botanische Studien (Jena)* 1: 1–338.

Vogel, S., Westerkamp, C., Thiel, B. and Gessner, K. (1984) Ornithophilie auf den Canarischen Inseln. *Plant Systematics and Evolution* 146: 225–248.

Williams, P.H. (1991) The bumble bees of the Kashmir Himalaya (Hymenoptera: Apidae, Bombini). *Bulletin of the British Museum* 60: 1–204.

Züchner, T., de Juana, E. and Boesman, P.F.D. (2020) Santa Marta Sabrewing (*Campylopterus phainopeplus*), version 1.0. *Birds of the World.* Cornell Lab of Ornithology, Ithaca, NY. https://doi.org/10.2173/bow.samsab1.01

Chapter 13 – The curious case of Europe

Burquez, A. (1989) Blue tits (*Parus caeruleus*) as pollinators of the crown imperial (*Fritillaria imperialis*) in Britain. *Oikos* 55: 335–340.

Coombs, G. and Peter, C.I. (2009) Do floral traits of *Strelitzia reginae* limit nectar theft by sunbirds? *South African Journal of Botany* 75: 751–756.

Darwin, C. (1874) Flowers of the primrose destroyed by birds. *Nature* 9: 482. Available at Darwin Online: http://darwin-online.org.uk

Darwin, C. (1876) *The Effects of Cross and Self Fertilisation in the Vegetable Kingdom*. John Murray, London.

Darwin, E. (1791) *The Botanic Garden. A poem in two parts. Part I containing the economy of vegetation. Part II. The loves of the plants. With philosophical notes.* – J. Johnson, St. Paul's Church Yard, London. Available at Project Gutenberg: www.gutenberg.org

da Silva, L.P., Ramos, J.A., Olesen, J.M., Traveset, A. and Heleno, R.H. (2014) Flower visitation by birds in Europe. *Oikos* 123: 1377–1383.

Ford, H.A. (1985) Nectarivory and pollination by birds in southern Australia and Europe. *Oikos* 44: 127–131.

Heard, S. (2023) How to use ChatGPT in scientific writing. https://scientistseessquirrel.wordpress.com/2023/06/20/how-to-use-chatgpt-in-scientific-writing

Kay, Q.O.N. (1985) Nectar from willow catkins as a food source for Blue Tits. *Bird Study* 32: 40–44.

Knuth, P. (1906–1909) *Handbook of Flower Pollination*. 3 vols. Clarendon Press, Oxford.

Lavaut, E., Guillemin, M.-L., Colin, S. *et al.* (2022) Pollinators of the sea: a discovery of animal-mediated fertilization in seaweed. *Science* 377: 528–530. See also the commentary by Ollerton and Ren, cited below.

Meeuse, B.J.D. (1962) *The Story of Pollination*. Ronald Press, New York.

Müller, H. (1873) *Die Befruchtung der Blumen durch Insekten, und die gegenseitigen Anpassungen beider*. Wilhelm Engelmann, Leipzig.

Müller, H. (1883) *The Fertilisation of Flowers*. Macmillan, London.

Navarro-Pérez, M.L., López, J., Fernández-Mazuecos, M. Rodríguez-Riaño, T., Vargas, P. and Ortega-Olivencia, A. (2013) The role of birds and insects in pollination shifts of *Scrophularia* (Scrophulariaceae). *Molecular Phylogenetics and Evolution* 69: 239–254.

Navarro-Pérez, M.L., López, J., Rodríguez-Riaño, T., Bacchetta, G., de Miguel Gordillo, C. and Ortega-Olivencia, A. (2017) Confirmed mixed bird–insect pollination system of *Scrophularia trifoliata* L., a Tyrrhenian species with corolla spots. *Plant Biology* 19: 460–468.

Ollerton, J. (2023) How reliable is ChatGPT? A weird encounter with Erasmus Darwin. https://jeffollerton.co.uk/2023/01/09/how-reliable-is-chatgpt-a-weird-encounter-with-erasmus-darwin

Ollerton, J. and Ren, Z.-X. (2022) Did pollination exist before plants? *Science* 377: 471–472.

Ortega-Olivencia, A., Rodríguez-Riaño, T., Valtueña, F.J., López, J. and Devesa, J.A. (2005) First confirmation of a native bird-pollinated plant in Europe. *Oikos* 110: 578–590.

Ortega-Olivencia, A., Rodríguez-Riaño, T., Pérez-Bote, J.L. *et al.* (2011) Insects, birds and lizards as pollinators of the largest-flowered *Scrophularia* of Europe and Macaronesia. *Annals of Botany* 109: 153–167. https://doi.org/10.1093/aob/mcr255

Peters, W.S., Pirl, M., Gottsberger, G. and Peters, D.S. (1995) Pollination of the Crown Imperial *Fritillaria imperialis* by Great Tits *Parus major*. *Journal of Ornithology* 136: 207–212.

White, G. (1789) *The Natural History and Antiquities of Selborne*. Benjamin White, London.

Wright, P. (1999) Obituary of Bastiaan Meeuse. *The Guardian*. www.theguardian.com/news/1999/aug/30/guardianobituaries1

Chapter 14 – 'After the Manner of Bees'

Ali, S.A. (1932) Flower-birds and bird-flowers in India. *Journal of the Bombay Natural History Society* 35: 573–605.

Anon. (2015) Contemplating nature: Chinese and Korean bird-and-flower paintings. https://eaa.fas.harvard.edu/contemplating-nature-chinese-and-korean-bird-and-flower-paintings

Avery, M. (2014) *A Message from Martha: The Extinction of the Passenger Pigeon and Its Relevance Today*. Bloomsbury, London.

Bailleul-LeSuer, R. (ed.) (2012) *Between Heaven and Earth: Birds in Ancient Egypt*. Oriental Institute Museum Publications, Chicago, IL. A copy is available here: https://oi-idb-static.uchicago.edu/multimedia/202/oimp35.pdf

Barlow, R. (1932) *A Brief Summe of Geographie*, ed. E.G.R. Taylor. Hakluyt Society, London. https://archive.org/details/dli.ministry.00999/page/n7/mode/2up

Catesby, M. (1729–1747) *The Natural History of Carolina, Florida and the Bahama Islands*. Privately published. www.biodiversitylibrary.org/item/126524#page/7/mode/1up

Darwin, C. (ed.) (1841) *Birds. Part 3 of The Zoology of the Voyage of H.M.S. Beagle by John Gould. Edited and superintended by Charles Darwin.* Smith Elder & Co., London.

de Léry, J. (1578) *Histoire d'un voyage fait en la Terre du Bresil, autrement dite Amerique.* Antoine Chuppin, La Rochelle. A copy is available here: https://purl.pt/136

de Oviedo, G.F. (1526) *De la natural hystoria de las Indias.* Facsimile edition, 1969. North Carolina Studies in the Romance Languages and Literatures. University of North Carolina Press, Chapel Hill, NC.

Docters van Leeuwen, W.M. (1954) On the biology of some Javanese Loranthaceae and the role birds play in their life-historie [*sic*]. *Beaufortia* 4: 103–207.

Doherty, J (T.) and Tumarae-Teka, K. (2015) Tūhoe Tuawhenua (Māori, New Zealand) knowledge of pollination and pollinators associated with food production. In Lyver, P., Perez, E., Carneiro da Cunha, M. and Roué, M. (eds), *Indigenous and Local Knowledge about Pollination and Pollinators Associated with Food Production: Outcomes from the Global Dialogue Workshop (Panama 1–5 December 2014).* UNESCO, Paris, pp. 27–37.

Eckert, S.L. and Clark, T. (2009) The ritual importance of birds in 14th-century Central New Mexico. *Journal of Ethnobiology* 29: 8–27.

Eda, M., Yamasaki, T. and Sakai, M. (2019) Identifying the bird figures of the Nasca pampas: an ornithological perspective. *Journal of Archaeological Science: Reports* 26: 101875. https://doi.org/10.1016/j.jasrep.2019.101875

Emeny, M.T., Powlesland, R.G., Henderson, I.M. and Fordham, R.A. (2009) Feeding ecology of Kererū (*Hemiphaga novaeseelandiae*) in podocarp–hardwood forest, Whirinaki Forest Park, New Zealand. *New Zealand Journal of Ecology* 33: 114–124.

Gibson Guitar Forum (2017) Hummingbird pickguard art. https://forum.gibson.com/topic/138933-hummingbird-pickguard-art

Grew, N. (1693) The description of the American tomineius, or hummingbird. *Philosophical Transactions* 17: 760–761.

Grew, N. (1693) A query put by Dr. N. Grew, concerning the food of the Humming Bird; occasioned by the description of it in the transactions. Numb. 200. *Philosophical Transactions* 17: 815.

Hooper, D. and Field, H. (1937) *Useful Plants and Drugs of Iran and Iraq.* Field Museum of Natural History: Botanical series publication 387. Field Museum, Chicago, IL. A copy can be found here: www.biodiversitylibrary.org/item/19714#page/3/mode/1up

Latham, J. (1790) *Index Ornithologicus, Sive Systema Ornithologiae: Complectens Avium Divisionem In Classes, Ordines, Genera, Species, Ipsarumque Varietates.* [2 Volumes, in Latin] Leigh and Sotheby, London. A copy can be found here: www.biodiversitylibrary.org/bibliography/131313

Lyver, P., Perez, E., Carneiro da Cunha, M. and Roué, M. (eds) (2015) *Indigenous and Local Knowledge about Pollination and Pollinators Associated with Food Production: Outcomes from the Global Dialogue Workshop (Panama 1–5 December 2014).* UNESCO, Paris.

Lyver, P.O'B., Taputu, M., Kutia, S.T. and Tahi, B. (2008) Tūhoe Tuawhenua mātauranga of Kererū (*Hemiphaga novaseelandiae novaseelandiae*) in Te Urewera. *New Zealand Journal of Ecology* 32: 7–17.

New Zealand Government (2022) Honey bees brought to New Zealand 19 March 1839. https://nzhistory.govt.nz/mary-bumby-brings-the-first-honey-bees-in-new-zealand.

Ortiz-Crespo, F.I. (1994) Sahagún's early hummingbird iconography and a review of the European discovery of trochilids. *Archives of Natural History* 21: 7–10.

Perrins, C. (1988) Obituary: Salim Moizuddin Abdul Ali. *Ibis* 130: 305–306.

PRS Guitars (2017) The story of the PRS bird inlays. https://prsguitars.com/blog/post/the_story_of_the_prs_bird_inlays

Roguz, K., Hill, L., Roguz, A. and Zych, M. (2021) Evolution of bird and insect flower traits in *Fritillaria* L. (Liliaceae). *Frontiers in Plant Science* 12: 656783.

Rumphius, G.E. (1741–1750) [Burmannus, J., ed.] *Het Amboinsche kruid-boek: Dat is, beschryving van de meest bekende boomen, heesters, kruiden, land- en water-planten, die men in Amboina.... Herbarium Amboinense, plurimas conplectens arbores, frutices, herbas, plantas terrestres & aquaticas, quae in Amboina...* (in Dutch and Latin). François Changuion, Jan Catuffe, Hermanus Uytwerf, Amsterdam.

Sahagún, B. de (1545–1590) *La Historia General de las Cosas de Nueva España.*

Sault, N. (2016) How hummingbird and vulture mediate between life and death in Latin America. *Journal of Ethnobiology* 36: 783–806.

Scott-Elliot, G.F. (1890) Ornithophilous flowers in South Africa. *Annals of Botany* 4: 265–280.

Shelley, G.E. (1876–1880) *A Monograph of the Nectariniidae, or Family of Sun-Birds.* Privately published, in parts. A copy is available here: www.biodiversitylibrary.org/bibliography/53516

Stimpson, C. and Kemp, B. (2023) Pigeons and papyrus at Amarna: the birds of the Green Room revisited. *Antiquity* 97: 104–119. https://doi.org/10.15184/aqy.2022.159. See also: www.livescience.com/realistic-bird-paintings-ancient-egypt

Stoudemire, S.A. (1959) *Translator and editor of Gonzalo Fernandez de Oviedo, Natural History of the West Indies.* University of North Carolina Studies in the Romance Languages and Literatures, No. 32. University of North Carolina Press, Chapel Hill, NC.

Taçon, P.S.C., Langley, M., May, S.K., Lamilami, R., Brennan, W. and Guse, D. (2010) Ancient bird stencils discovered in Arnhem Land, Northern Territory, Australia. *Antiquity* 84: 416–427.

Thevet, A. (1557) *Les Singularitez de la France Antarctique.* Published in English in 1568 as *The New Found Worlde, or Antarctike, wherein is contained wōderful and strange things, as well of humaine creatures, as Beastes, Fishes, Foules, and Serpents, Trees, Plants, Mines of Golde and Siluer.*

Vigors, N.A. (1825) Observations on the natural affinities that connect the orders and families of birds. *Transactions of the Linnean Society of London* 14: 395–516. A copy can be found here: www.biodiversitylibrary.org/bibliography/49426

Wali, A. and Odland, J.C. (eds) (2016) The Shipibo-Conibo: culture and collections in context. *Fieldiana Anthropology* 45: 1–100.

Wei, J., Huo, Z., Gorb, S.N., Rico-Guevara, A., Wu, Z. and Wu, J. (2020) Sucking or lapping: facultative feeding mechanisms in honeybees (*Apis mellifera*). *Biological Letters* 16: 20200449. https://doi.org/10.1098/rsbl.2020.0449. See also: www.independent.co.uk/climate-change/news/honeybee-nectar-consumption-sucking-dipping-tongues-discovery-a9666346.html

Chapter 15 – Feathers and fruits

Bailes, E., Ollerton, J., Pattrick, J. and Glover, B.J. (2015) How can an understanding of plant–pollinator interactions contribute to global food security? *Current Opinion in Plant Biology* 26: 72–79.

Brittain, C., Kremen, C., Garber, A. and Klein, A.-M. (2014) Pollination and plant resources change the nutritional quality of almonds for human health. *PLoS One* 9: e90082. https://doi.org/10.1371/journal.pone.0090082

Chaplin-Kramer, R., Dombeck, E., Gerber, J. *et al.* (2014) Global malnutrition overlaps with pollinator-dependent micronutrient production. *Proceedings of the Royal Society B* 281: 20141799. https://doi.org/10.1098/rspb.2014.1799

Ducroquet, J.P.H.J. and Hickel, E.R. (1997) Birds as pollinators of feijoa (*Acca sellowiana* Bert). *Acta Horticulturae* 452: 37–40.

Egerer, M.H., Fricke, E.C. and Rogers, H.S. (2018) Seed dispersal as an ecosystem service: frugivore loss leads to decline of a socially valued plant, *Capsicum frutescens*. *Ecological Applications* 28: 655–667.

Eilers, E.J., Kremen, C., Greenleaf, S.S., Garber, A.K. and Klein, A.-M. (2011) Contribution of pollinator-mediated crops to nutrients in the human food supply. *PLoS One* 6: e21363. https://doi.org/10.1371/journal.pone.0021363

Ellis, A.M., Myers, S.S. and Ricketts, T.H. (2015) Do pollinators contribute to nutritional health? *PLoS One* 10: e114805. https://doi.org/10.1371/journal.pone.0114805

Fang, Q., Chen, Y.-Z. and Huang S.-Q. (2012) Generalist passerine pollination of a winter-flowering fruit tree in central China. *Annals of Botany* 109: 379–384.

Greenberg, R., Bichier, P., Angon, A.C., MacVean, C., Perez, R. and Cano, E. (2000) The impact of avian insectivory on arthropods and leaf damage in some Guatemalan coffee plantations. *Ecology* 81: 1750–1755.

Holmes, R.J. and Froud-Williams, R.J. (2005) Post-dispersal weed seed predation by avian and non-avian predators. *Agriculture, Ecosystems and Environment* 105: 23–27.

Itino, T., Kato, M. and Hotta, M. (1991) Pollination ecology of the two wild bananas, *Musa acuminata* subsp. *halabanensis* and *M. salaccensis*: chiropterophily and ornithophily. *Biotropica* 23: 151–158.

Kamala Jayanthi, P.D., Murthy, B.N.S., Nagaraja, T. *et al.* (2015) Nectar robbing by Purple Sunbird, *Nectarinia asiatica* (L.) in Pomegranate reduces reproductive success. *Pest Management in Horticultural Ecosystems* 21: 214–218.

Karp, D.S., Mendenhall, C.D., Sandí, R.F. *et al.* (2013) Forest bolsters bird abundance, pest control and coffee yield. *Ecology Letters* 16: 1339–1347.

Kellermann, J.L., Johnson, M.D., Stercho, A.M. and Hackett, S.C. (2008) Ecological and economic services provided by birds on Jamaican Blue Mountain coffee farms. *Conservation Biology* 22: 1177–1185.

Klein, A.-M., Vaissière, B.E., Cane, J.H. *et al.* (2007) Importance of pollinators in changing landscapes for world crops. *Proceedings of the Royal Society of London B* 274: 303–313.

Kwapong, P. and Kudom, A. (2010) Floral visitors of *Ananas comosus* in Ghana: a preliminary assessment. *Journal of Pollination Ecology* 2: 27–32.

Liu, A.Z., Li, D.Z., Wang, H. and Kress, W.J. (2002) Ornithophilous and chiropterophilous pollination in *Musa itinerans* (Musaceae), a pioneer species in tropical rain forests of Yunnan, southwestern China. *Biotropica* 34: 254–260.

Matallana-Puerto, C.A. and Cardoso, J.C.F. (2022) Ratatouille of flowers! Rats as potential pollinators of a petal-rewarding plant in the urban area. *Ecology* 103: e3778. https://doi.org/10.1002/ecy.3778.

Maués, M.M and Venturieri, G.C (1997) Pollination ecology of *Platonia insignis* Mart. (Clusiaceae), a fruit tree from eastern Amazon region. *Acta Horticulturae* 437: 255–260.

Medina, G. and Ferreira, S. (2004) Bacuri (*Platonia insignis* Martius), the Amazonian fruit that has become gold. In Alexiades, M.N. and Shanley, P. (eds), *Forest Products, Livelihoods and Conservation: Case Studies of Non-Timber Forest Product Systems.* Center for International Forestry Research, Bogor, Indonesia, pp. 195–210. www.jstor.org/stable/resrep02086.15

Moo-Aldana, R.D., Munguía-Rosas, M.A., Serralta, L.P., Castillo-Burguete, M.T., Vega-Frutis, R. and Martínez-Natarén, D. (2017) Can the introduction of modern crop varieties in their centre of origin affect local ecological knowledge? A case study of papaya in the Yucatan Peninsula. *Human Ecology* 45: 367–375.

Nadra, M.G., Giannini, N.P., Acosta, J.M. and Aagesen, L. (2018) Evolution of pollination by frugivorous birds in Neotropical Myrtaceae. *PeerJ* 6: e5426. https://doi.org/10.7717/peerj.5426

Nature in Bottle (2023) Bacuri Oil promotion. www.natureinbottle.com/product/bacuri_oil

Nature Observer (2022) Purple-rumped Sunbird taking nectar from pomegranate flowers. www.youtube.com/watch?v=ojsDJHCwojU

Nur, N. (1976) Studies on pollination in Musaceae. *Annals of Botany* 40: 167–177.

Nyffeler, M., Şekercioğlu, Ç.H. and Whelan, C.J. (2018) Insectivorous birds consume an estimated 400–500 million tons of prey annually. *Science of Nature* 105: 47. https://doi.org/10.1007/s00114-018-1571-z

Ollerton, J., Rouquette, J.R. and Breeze, T.D. (2016) Insect pollinators boost the market price of culturally important crops: holly, mistletoe and the spirit of Christmas. *Journal of Pollination Ecology* 19: 93–97.

Prendergast, K. (2023) YouTube channel. www.youtube.com/c/TheBeeBabette.

Rader, R., Bartomeus, I., Garibaldi, L.A. *et al.* (2016) Non-bee insects are important contributors to global crop pollination. *Proceedings of the National Academy of Sciences of the USA* 113: 146–151.

Ramíreza, F. and Kallarackal, J. (2017) Feijoa [*Acca sellowiana* (O. Berg) Burret] pollination: A review. *Scientia Horticulturae* 226: 333–341.

Roubik, D.W. (ed.) (1995) *Pollination of Cultivated Plants in the Tropics.* FAO Agricultural Services Bulletin 118. FAO, Rome.

Sabino, W., Costa, L., Andrade, T. *et al.* (2022) Status and trends of pollination services in Amazon agroforestry systems. *Agriculture, Ecosystems and Environment*. 335: 108012. https://doi.org/10.1016/j.agee.2022.108012

Sazima, I. and Sazima, M. (2007) Floral titbits: petals of *Acca sellowiana* (Myrtaceae) as a food source for birds in an urban area in southern Brazil. *Biota Neotropica* 7: https://doi.org/10.1590/S1676-06032007000200035

Smith, M.R., Singh, G.M., Mozaffarian, D. and Myers, S.S. (2015) Effects of decreases of animal pollinators on human nutrition and global health: a modelling analysis. *Lancet* 386: 1964–1972. https://doi.org/10.1016/S0140-6736(15)61085-6

Stewart, A.M. and Craig, J.L. (1989) Factors affecting pollinator effectiveness in *Feijoa sellowiana. New Zealand Journal of Crop and Horticultural Science* 17: 145–154.

Sun, S.-G., Huang, Z.-H., Chen, Z.-B. and Huang, S.-Q. (2017) Nectar properties and the role of sunbirds as pollinators of the golden-flowered tea (*Camellia petelotii*). *American Journal of Botany* 104: 468–476.

US National Park Service (2015) Saguaro fruit: a traditional harvest. www.nps.gov/sagu/learn/historyculture/upload/Saguaro-Fruit-A-Traditional-Harvest-Brief.pdf

Westerkamp, C. and Gottsberger, G. (2000) Diversity pays in crop pollination. *Crop Science* 40: 1209–1222.

Whelan, C.J., Wenny, D.G. and Marquis, R.J. (2008) Ecosystem services provided by birds. *Annals of the New York Academy of Sciences* 1134: 25–60.

Wolowski, M., Agostini, K., Rech, A.R. *et al.* (2019) *Relatório temático sobre polinização, polinizadores e produção de alimentos no Brasil.* REBIPP, Espiríto Santo. www.bpbes.net.br/wp-content/uploads/2019/03/BPBES_CompletoPolinizacao-2.pdf

Zurek, M., Ingram, J., Bellamy, A.S. *et al.* (2022) Food system resilience: concepts, issues and challenges. *Annual Review of Environment and Resources* 47: 511–534.

Chapter 16 – Urban flowers for urban birds

Anselmo, P.A., Cardoso, J.C.F., Siqueira, P.R. and Maruyama, P.K. (2023) Non-native plants and illegitimate interactions are highly relevant for supporting hummingbird pollinators in the urban environment. *Urban Forestry and Urban Greening* 86: 128025. https://doi.org/10.1016/j.ufug.2023.128025

Arizmendi, M.C., Constanza, M.S., Lourdes, J., Ivonne, I.F.M. and Edgar, L.S. (2007) Effect of the presence of nectar feeders on the breeding success of *Salvia mexicana* and *Salvia fulgens* in a suburban park near Mexico City. *Biological Conservation* 136: 155–158.

Aronson, M.F.J., La Sorte, F.A., Nilon, C.H. *et al.* (2014) A global analysis of the impacts of urbanization on bird and plant diversity reveals key anthropogenic drivers. *Proceedings of the Royal Society B* 281: 20133330. https://doi.org/10.1098/rspb.2013.3330

Coetzee, A., Barnard, P. and Pauw, A. (2018) Urban nectarivorous bird communities in Cape Town, South Africa, are structured by ecological generalisation and resource distribution. *Journal of Avian Biology* 49: jav-01526. https://doi.org/10.1111/jav.01526

Cortés, J.E. (1982) Nectar feeding by European passerines on introduced tropical flowers in Gibraltar. *Alectoris* 4: 26–29.

du Plessis, M., Seymour, C.L., Spottiswoode, C.N. and Coetzee, A. (2021) Artificial nectar feeders reduce sunbird abundance and plant visitation in Cape Fynbos adjacent to suburban areas. *Global Ecology and Conservation* 28: e01706. https://doi.org/10.1016/j.gecco.2021.e01706

French, K., Major, R. and Hely, K. (2005) Use of native and exotic garden plants by suburban nectarivorous birds. *Biological Conservation* 121: 545–559.

Hall, D.M., Camilo, G.D., Tonietto, R.K. *et al.* (2017) The city as a refuge for insect pollinators. *Conservation Biology* 31: 24–29.

Hughes, A.C., Orr, M.C., Lei, F., Yang, Q. and Qiao, H. (2022) Understanding drivers of global urban bird diversity. *Global Environmental Change* 76: 102588. https://doi.org/10.1016/j.gloenvcha.2022.102588

Inouye, D.W., Calder, W.A. and Waser, N.M. (1991) The effect of floral abundance on feeder censuses of hummingbird populations. *The Condor* 93: 279–285.

Leveau, L.M. (2008) Dynamics of nectarivory in the house sparrow in an urban environment. *Ornitologia Neotropical* 19: 275–281.

Lim, B.T.M., Sadanandan, K.R., Dingle, C. *et al.* (2019) Molecular evidence suggests radical revision of species limits in the great speciator white-eye genus *Zosterops*. *Journal of Ornithology* 160: 1–16.

Liu, Z., He, C., Zhou, Y. and Wu, J. (2014) How much of the world's land has been urbanized, really? A hierarchical framework for avoiding confusion. *Landscape Ecology* 29: 763–771.

Marín-Gómez, O.H., Flores, C.R. and del Coro Arizmendi, M. (2022) Assessing ecological interactions in urban areas using citizen science data: insights from hummingbird–plant meta-networks in a tropical megacity. *Urban Forestry and Urban Greening* 74: 127658. https://doi.org/10.1016/j.ufug.2022.127658

Maruyama, P.K. Bonizário, C., Marcon, A.P. *et al.* (2019) Plant–hummingbird interaction networks in urban areas: generalization and the importance of trees with specialized flowers as a nectar resource for pollinator conservation. *Biological Conservation* 230: 187–194.

Mendonça, L.B. and dos Anjos, L. (2006) Feeding behavior of humming hummingbirds and perching birds on *Erythrina speciosa* Andrews (Fabaceae) flowers in an urban area, Londrina, Paraná, Brazil. *Revista Brasileira de Zoologia* 23: 42–49.

Nakamura, V.A., Souza, C.S. and Araujo, A.C. (2023) Mass-flowering native species are key in the structure of an urban plant–hummingbird network. *Urban Ecosystems* 26: 929–940. https://doi.org/10.1007/s11252-023-01346-8

Ollerton, J., Trunschke, J., Havens, K. *et al.* (2022) Pollinator–flower interactions in gardens during the COVID-19 pandemic lockdown of 2020. *Journal of Pollination Ecology* 31: 87–96. https://doi.org/10.26786/1920-7603(2022)695

Ortega-Olivencia, A., Rodríguez-Riaño, T., Valtueña, F.J., López, J. and Devesa, J.A. (2005) First confirmation of a native bird-pollinated plant in Europe. *Oikos* 110: 578–590.

Pauw, A. and Louw, K. (2012) Urbanization drives a reduction in functional diversity in a guild of nectar-feeding birds. *Ecology and Society* 17: 27. https://doi.org/10.5751/ES-04758-170227

Previatto, D., Mizobe, R. and Posso, S. (2013) Birds as potential pollinators of the *Spathodea nilotica* (Bignoniaceae) in the urban environment. *Brazilian Journal of Biology* 73: 737–741.

Puga-Caballero, A., Arizmendi, M.D. and Sánchez-González, L.A. (2020) Phylogenetic and phenotypic filtering in hummingbirds from urban environments in central Mexico. *Evolutionary Ecology* 34: 525–541.

Rodrigues, L. and Araujo, A. (2017) The hummingbird community and their floral resources in an urban forest remnant in Brazil. *Brazilian Journal of Biology* 71: 611–622.

Sazima, I and Sazima, M. (2022) A handful of beauty and services: flower-visiting birds at two small urbanised sites in south-eastern Brazil and Australia. *Flora* 296: 152151. https://doi.org/10.1016/j.flora.2022.152151

Silva, P.A., Silva, L.L. and Brito, L. (2020) Using bird–flower interactions to select native tree resources for urban afforestation: the case of *Erythrina velutina*. *Urban Forestry and Urban Greening* 51: 126677. https://doi.org/10.1016/j.ufug.2020.126677

Silva, V.H.D., Gomes, I.N.G., Cardoso, J.C.F. *et al.* (2023) Diverse urban pollinators and where to find them. *Biological Conservation* 281: 110036. https://doi.org/10.1016/j.biocon.2023.110036

Sonne, J., Kyvsgaard, P., Maruyama, P.K. *et al.* (2016) Spatial effects of artificial feeders on hummingbird abundance, floral visitation and pollen deposition. *Journal of Ornithology* 157: 573–581.

UNCTAD (2022) Handbook of Statistics: Total and urban population. https://hbs.unctad.org/total-and-urban-population.

Vitali-Veiga, M.J. and Machado, V.L.L. (2000) Visitantes florais de *Erythrina speciosa* Andr. (Leguminosae). *Revista Brasileira de Zoologia* 17: 369–383.

Vitorino, B.D., da Frota, A.V.B. and Maruyama, P.K. (2021) Ecological determinants of interactions as key when planning pollinator-friendly urban greening: a plant–hummingbird network example. *Urban Forestry and Urban Greening* 64: 127298. https://doi.org/10.1016/j.ufug.2021.127298

Vonnegut, K. (1969) *Slaughterhouse Five.* Delacorte, New York.

Wolf, N., Smeltz, T.S., Cook, C. Martinez del Rio, C. (2023) Using stable isotopes in hummingbird breath to estimate reliance on supplemental feeders. *Ecology and Evolution* 3: e9799. https://doi.org/10.1002/ece3.9799

Chapter 17 – Bad birds and feral flowers

Chandar, G.S. (2010) Floral phenology, pollination and fruiting patterns in *Delonix regia* and *D. elata* (Caesalpiniaceae). Doctoral dissertation, Andhra University, India.

Costa, A., Moré, M., Sérsic, A.N. *et al.* (2023) Floral colour variation of *Nicotiana glauca* in native and non-native ranges: Testing the role of pollinators' perception and abiotic factors. *Plant Biology* 25: 403–410. https://doi.org/10.1111/plb.13509

Crates, R., McDonald, P.G., Melton, C.B. *et al.* (2023) Towards effective management of an overabundant native bird: the noisy miner. *Conservation Science and Practice* 5: e12875. https://doi.org/10.1111/csp2.12875

Diamond, J.M. (1974) Colonization of exploded volcanic islands by birds: the supertramp strategy. *Science* 184: 803–806.

Diamond, J.M., Pimm, S.L., Gilpin, M.E. and LeCroy, M. (1989) Rapid evolution of character displacement in myzomelid honeyeaters. *The American Naturalist* 134: 675–708.

Florentine, S.K., Westbrooke, M.E., Gosney, K., Ambrose, G. and O'Keefe, M. (2006) The arid land invasive weed *Nicotiana glauca* R. Graham (Solanaceae): population and soil seed bank dynamics, seed germination patterns and seedling response to flood and drought. *Journal of Arid Environments* 66: 218–230.

Fountain, J. and McDonald, P.G. (2022) Do differences in the availability of anthropogenic food resources influence the observed levels of agonistic behaviour in Noisy Miners (*Manorina melanocephala*)? *Emu – Austral Ornithology* 122: 61–70. See also: https://theconversation.com/want-noisy-miners-to-be-less-despotic-think-twice-before-filling-your-garden-with-nectar-rich-flowers-190226

Freed, L.A. and Cann, R.L. (2009) Negative effects of an introduced bird species on growth and survival in a native bird community. *Current Biology* 19: 1736–1740.

Global Invasive Species Database (2023) 100 of the world's worst invasive alien species. www.iucngisd.org/gisd/100_worst.php

Global Invasive Species Database (2023) Species profile: *Nicotiana glauca*. www.iucngisd.org/gisd/speciesname/Nicotiana+glauca

Grey, M.J., Clarke, M.F. and Loyn, R.H. (1997) Initial changes in avian communities of remnant eucalypt woodlands following a reduction in the abundance of noisy miners, *Manorina melanocephala*. *Wildlife Research* 24: 631–648.

Grey, M.J., Clarke, M.F. and Loyn, R.H. (1998) Influence of the noisy miner *Manorina melanocephala* on avian diversity and abundance in remnant grey box woodland. *Pacific Conservation Biology* 4: 55–69.

Gunasekaran, M., Trabelcy, B., Izhaki, I., Halpern, M. (2021) Direct evidence that sunbirds' gut microbiota degrades floral nectar's toxic alkaloids. *Frontiers in Microbiology* 12. https://doi.org/10.3389/fmicb.2021.639808

Huxley, E. (1959) *The Flame Trees of Thika: Memories of an African Childhood.* W. Morrow, London.

Invasives of South Africa (2023) https://invasives.org.za/fact-sheet/wild-tobacco

Kessler, D., Bhattacharya, S., Diezel, C. *et al.* (2012) Unpredictability of nectar nicotine promotes outcrossing by hummingbirds in *Nicotiana attenuata. The Plant Journal* 71: 529–538.

Lammers, T.G., Weller, S.G. and Sakai AK. (1987) Japanese White-eye, an introduced passerine, visits the flowers of *Clermontia arhorescens*, an endemic Hawaiian lobelioid. *Pacific Science* 41: 1–4.

Leven, M.R. and Corlett, R.T. (2004) Invasive birds in Hong Kong, China. *Ornithological Science* 3: 43–55.

Lewis, G. (2020) *Delonix regia. Curtis's Botanical Magazine* 37: 324–331.

Mackin, C.R., Peña, J.F., Blanco, M.A., Balfour, N.J. and Castellanos, M.C. (2021) Rapid evolution of a floral trait following acquisition of novel pollinators. *Journal of Ecology* 109: 2234–2246.

Mandelbaum, R.F. (2023) Parrot invasions. *Scientific American* 329: 40–49. www.scientificamerican.com/article/parrots-are-taking-over-the-world

Maruyama, P.K., Vizentin-Bugoni, J., Sonne, J. *et al.* (2016) The integration of alien plants in plant-hummingbird pollination networks across the Americas: the importance of species traits and insularity. *Diversity and Distributions* 22: 672–681.

Memmott, J. and Waser, N.M. (2002) Integration of alien plants into a native flower–pollinator visitation web. *Proceedings of the Royal Society B* 269: 2395–2399.

Moles, A.T., Dalrymple, R.L., Raghu, S., Bonser, S.P. and Ollerton, J. (2022) Advancing the missed mutualist hypothesis, the under-appreciated twin of the enemy release hypothesis. *Biology Letters* 18: 20220220. https://doi.org/10.1098/rsbl.2022.0220

Norman, J. and Christidis, L. (2014) Invasion ecology of honeyeaters. In Prins, H. and Gordon, I. (eds), *Invasion Biology and Ecological Theory: Insights from a Continent in Transformation*. Cambridge University Press, Cambridge, pp. 83–102.

Ollerton, J., Watts, S., Connerty, S. *et al.* (2012) Pollination ecology of the invasive tree tobacco *Nicotiana glauca*: comparisons across native and non-native ranges. *Journal of Pollination Ecology* 9: 85–95.

Pimenta, V.R.A., Dias, M.M. and Reis, M.G. (2020) Hummingbird (Aves: Trochilidae) assemblage using resources from the exotic African tuliptree, *Spathodea campanulata* (Bignoniaceae) in a Neotropical altered environment, southeastern Brazil. *Brazilian Journal of Biology* 81: 137–143.

Queiroz, A.C.M., Contrera, F.A.L. and Venturieri, G.C. (2014) The effect of toxic nectar and pollen from *Spathodea campanulata* on the worker survival of *Melipona fasciculata* Smith and *Melipona seminigra* Friese, two Amazonian stingless bees (Hymenoptera: Apidae: Meliponini). *Sociobiology* 61: 536–540.

Rangaiah, K., Purnachandra Rao, S. and Solomon Raju, A.J. (2004) Bird-pollination and fruiting phenology in *Spathodea campanulata* Beauv. (Bignoniaceae). *Beitrage zur Biologie der Pflanzen* 73: 395–408.

Richardson, D.M., Allsopp, N., D'Antonio, C.M., Milton, S.J. and Rejmanek, M. (2000) Plant invasions; the role of mutualisms. *Biological Reviews* 75: 65–93.

Roy, D., Alderman, D., Anastasiu, P. *et al.* (2020) DAISIE – Inventory of alien invasive species in Europe. Version 1.7. Research Institute for Nature and Forest (INBO). Checklist dataset. https://doi.org/10.15468/ybwd3x

Schueller, S.K. (2004) Self-pollination in island and mainland populations of the introduced hummingbird-pollinated plant, *Nicotiana glauca* (Solanaceae). *American Journal of Botany* 91: 672–681.

Tadmor-Melamed, H., Markman, S., Arieli, A., Distl, M., Wink, M. and Izhaki, I. (2004) Limited ability of Palestine Sunbirds *Nectarinia osea* to cope with pyridine alkaloids in nectar of Tree Tobacco *Nicotiana glauca*. *Functional Ecology* 18: 844–850.

US Forest Service (2013) Pacific Island Ecosystems at Risk (PIER). www.hear.org/pier/species/nicotiana_glauca.htm

Yap, C.A. and Sodhi, N.S. (2004) Southeast Asian invasive birds: ecology, impact and management. *Ornithological Science* 3: 57–67.

Chapter 18 – What escapes the eye

Ali, J.R., Blonder, B.W., Pigot, A.L. and Tobias, J.A. (2023) Bird extinctions threaten to cause disproportionate reductions of functional diversity and uniqueness. *Functional Ecology* 37: 162–175. https://doi.org/10.1111/1365-2435.14201. See also: https://phys.org/news/2022-11-planet-unique-birds-higher-extinction.html

Anderson, S.H., Kelly, D., Ladley, J.L., Molloy, S. and Terry, J. (2011) Cascading effects of bird functional extinction reduce pollination and plant density. *Science* 331: 1068–1071.

Aslan, C.E., Zavaleta, E.S., Tershy, B., Croll, D. and Robichaux, R.H. (2014) Imperfect replacement of native species by non-native species as pollinators of endemic Hawaiian plants. *Conservation Biology* 28: 478–488.

Aslan, C.E., Liang, C.T., Shiels, A.B. and Haines, W. (2018) Absence of native flower visitors for the endangered Hawaiian mint *Stenogyne angustifolia*: impending ecological extinction? *Global Ecology and Conservation* 16: e00468. https://doi.org/10.1016/j.gecco.2018.e00468

Berryman, A.J., Collar, N.J., Crozariol, M.A., Gussoni, C.O.A., Kirwan, G.M. and Sharpe, C.J. (2023) The distribution, ecology and conservation status of the long-tailed woodnymph *Thalurania watertonii*. *Ornithological Research* 31: 1–12. https://doi.org/10.1007/s43388-022-00110-4

BirdLife International (2022) *State of the World's Birds: 2022 Annual Update*. http://datazone.birdlife.org/2022-annual-update.

Boyd, R.J., Aizen, M.A., Barahona-Segovia, R.M. *et al.* (2022) Inferring trends in pollinator distributions across the Neotropics from publicly available data remains challenging despite mobilization efforts. *Diversity and Distributions* 28: 1404–1415. https://doi.org/10.1111/ddi.13551

Carpenter, R.J., Jordan, G.J. and Hill, R.S. (1994) *Banksieaephyllum taylorii* (Proteaceae) from the late Paleocene of New South Wales and its relevance to the origin of Australia's scleromorphic flora. *Australian Systematic Botany* 7: 385–392. https://doi.org/10.1071/SB9940385

Carpenter, R.J., Jordan, G.J. and Hill, R.S. (2016) Fossil leaves of *Banksia*, Banksieae and pretenders: resolving the fossil genus *Banksieaephyllum*. *Australian Systematic Botany* 29: 126–141. https://doi.org/10.1071/SB16005

Cox, P.A. (1983) Extinction of the Hawaiian avifauna resulted in a change of pollinators for the ieie, *Freycinetia arborea. Oikos* 41: 195–199.

Crates, R., Langmore, N., Ranjard, L. *et al.* (2021) Loss of vocal culture and fitness costs in a critically endangered songbird. *Proceedings of the Royal Society B* 288: 20210225. https://doi.org/10.1098/rspb.2021.0225

Eaton, J.A., Shepherd, C.R., Rheindt, F.E. *et al.* (2015) Trade-driven extinctions and near-extinctions of avian taxa in Sundaic Indonesia. *Forktail* 31: 1–12.

Geerts, S. (2016) Can short-billed nectar thieving sunbirds replace long-billed sunbird pollinators in transformed landscapes? *Plant Biology* 18: 1048–1052.

Geyle, H.M., Woinarski, J.C.Z., Baker, G.B. *et al.* (2018) Quantifying extinction risk and forecasting the number of impending Australian bird and mammal extinctions. *Pacific Conservation Biology* 24: 157–167.

Global Biodiversity Information Facility [GBIF] (2023) www.gbif.org.

Greig, E.I., Wood, E.M. and Bonter, D.N. (2017) Winter range expansion of a hummingbird is associated with urbanization and supplementary feeding. *Proceedings of the Royal Society B* 284: 20170256. https://doi.org/10.1098/rspb.2017.0256

Higgins, P.J., Christidis, L. and Ford, H. (2020) Scarlet Myzomela (*Myzomela sanguinolenta*), version 1.0. *Birds of the World*. Cornell Lab of Ornithology, Ithaca, NY. https://doi.org/10.2173/bow.scamyz1.01

Janzen, D.H. (1974) The deflowering of central America. *Natural History* 83: 49–53.

Kelly, D., Ladley, J.J., Robertson, A.W., Anderson, S.H., Wotton, D.M. and Wiser, S.K. (2010) Mutualisms with the wreckage of an avifauna: the status of bird pollination and fruit-dispersal in New Zealand. *New Zealand Journal of Ecology* 34: 66–85.

McNamara, K.J. and Scott, J.K. (2008) A new species of *Banksia* (Proteaceae) from the Eocene Merlinleigh Sandstone of the Kennedy Range, Western Australia. Alcheringa. 7: 185–193.

Park, G. (2022) The scarlet myzomela. https://geoffpark.wordpress.com/2022/09/12/the-scarlet-myzomela

Paton, D.C. (1993) Honeybees in the Australian environment. *BioScience* 43: 95–103.

Pender, R.J., Morden, C.W. and Paull, R.E. (2014) Investigating the pollination syndrome of the Hawaiian lobeliad genus *Clermontia* (Campanulaceae) using floral nectar traits. *American Journal of Botany* 101: 201–205. https://doi.org/10.3732/ajb.1300338

Säterberg, T., Sellman, S. and Ebenman, B. (2013) High frequency of functional extinctions in ecological networks. *Nature* 499: 468–470. https://doi.org/10.1038/nature12277

Smith, T.B., Freed, L.A., Lepson, J.K. and Carothers, J.H. (1995) Evolutionary consequences of extinctions in populations of a Hawaiian honeycreeper. *Conservation Biology* 9: 107–113.

Sonne, J., Maruyama, P.K., Martín González, A.M., Rahbek, C., Bascompte, J. and Dalsgaard, B. (2022) Extinction, coextinction and colonization dynamics in plant–hummingbird networks under climate change. *Nature Ecology & Evolution* 6: 720–729. https://doi.org/10.1038/s41559-022-01693-3

Winkler, D.W., Billerman, S.M. and Lovette, I.J. (2020) Honeyeaters (Meliphagidae), version 1.0. *Birds of the World*. Cornell Lab of Ornithology, Ithaca, NY. https://doi.org/10.2173/bow.meliph3.01

Chapter 19 – The restoration of hope

Atkinson, C.T., Saili, K.S., Utzurrum, R.B. and Jarvi S.I. (2013) Experimental evidence for evolved tolerance to avian malaria in a wild population of low elevation Hawai'i 'Amakihi (*Hemignathus virens*). *EcoHealth* 10: 366–375.

Bush, M.B., Rozas-Davila, A., Raczka, M. *et al.* (2022) A palaeoecological perspective on the transformation of the tropical Andes by early human activity. *Philosophical Transactions of the Royal Society B* 377: 20200497. https://doi.org/10.1098/rstb.2020.0497. See also the other articles in the theme issue 'Tropical forests in the deep human past': https://royalsocietypublishing.org/toc/rstb/2022/377/1849

Carey, P. (2020) The best strategy for using trees to improve climate and ecosystems? Go natural. *Proceedings of the National Academy of Sciences of the USA* 117: 4434–4438. https://doi.org/10.1073/pnas.2000425117

Cruz-Neto, O., Souza e Silva, J.L., Woolley, M.M. Tabarelli, M. and Lopes, A.V. (2018) Pollination partial recovery across monospecific plantations of a native tree (*Inga vera*, Leguminosae) in the Atlantic forest: lessons for restoration. *Forest Ecology and Management* 427: 383–391.

Dalsgaard, B., Magård, E., Fjeldså, J. *et al.* (2011) Specialization in plant–hummingbird networks is associated with species richness, contemporary precipitation and Quaternary climate-change velocity. *PLoS One* 6: e25891. https://doi.org/10.1371/journal.pone.0025891

Freed, L.A., Cann, R.L., Goff, M.L. Kuntz, W.A. and Bodner, G.R. (2005) Increase in avian malaria at upper elevation in Hawai'i. *The Condor* 107: 753–764.

Frick, K.M., Ritchie, A.L. and Krauss, S.L. (2014) Field of dreams: restitution of pollinator services in restored bird-pollinated plant populations. *Restoration Ecology* 22: 832–840.

Genes, L. and Dirzo, R. (2022) Restoration of plant-animal interactions in terrestrial ecosystems. *Biological Conservation* 265: 109393. https://doi.org/10.1016/j.biocon.2021.109393

Gimesy, D. (2021) The road to saving Australia's regent honeyeaters. www.australiangeographic.com.au/topics/wildlife/2021/01/the-road-to-saving-australias-regent-honeyeaters. See also: www.youtube.com/watch?v=9Nw1nIP7Nco

Hawaii Department of Land and Natural Resources (2022) RACE TO SAVE HAWAIIAN HONEYCREEPERS BOLSTERED BY $14 MILLION IN FEDERAL INFRASTRUCTURE AID. https://dlnr.hawaii.gov/blog/2022/05/16/nr22-068

Ingcungcu Sunbird Restoration Project (2023) https://ingcungcu.org

Island Conservation (2022) Locally extinct birds and geckos reappear on Pinzón and Rábida Islands, Galapagos ten years after rodent removal. www.islandconservation.org/press-release-locally-extinct-birds-and-geckos-reappear-on-pinzon-and-rabida-islands-galapagos-ten-years-after-rodent-removal

Kaiser-Bunbury, C.N., Memmott J. and Müller C.B. (2009) Community structure of pollination webs of Mauritian heathland habitats. *Perspectives in Plant Ecology, Evolution and Systematics* 11: 241–254.

Kaiser-Bunbury, C.N., Muff, S., Memmot, J., Muller, C.B. and Caflisch, A. (2010) The robustness of pollination networks to the loss of species and interactions: a quantitative approach incorporating pollinator behaviour. *Ecology Letters* 13: 442–452.

Kaiser-Bunbury, C., Mougal, J., Whittington, A. *et al.* (2017) Ecosystem restoration strengthens pollination network resilience and function. *Nature* 542: 223–227.

Lindell, C.A. and Thurston, G.M. (2013) Bird pollinator visitation is equivalent in island and plantation planting designs in tropical forest restoration sites. *Sustainability* 5: 1177–1187.

Lindsey, G.D., VanderWerf, E.A., Baker, H. and Baker, P.E. (2020) Hawaii Amakihi (*Chlorodrepanis virens*), version 1.0. *Birds of the World*. Cornell Lab of Ornithology, Ithaca, NY.

Mann, C.C. (2005) *1491: New Revelations of the Americas Before Columbus*. Knopf, New York. Published in the UK as *Ancient Americans: Rewriting the History of the New World*.

Mann, M.E. (2021) *The New Climate War: the Fight to Take Back Our Planet*. Scribe Publications, London.

Mann, M.E., Bradley, R.S. and Hughes, M.K. (1999) Northern hemisphere temperatures during the past millennium: inferences, uncertainties, and limitations. *Geophysical Research Letters* 26: 759–762.

Maui Forest Bird Recovery Project (2023) www.mauiforestbirds.org/forest-restoration

Mnisi, B.E., Geerts, S., Smith, C. and Pauw, A. (2021) Nectar gardens on school grounds reconnect plants, birds and people. *Biological Conservation* 257: 109087. https://doi.org/10.1016/j.biocon.2021.109087

Poorter, L., Craven, D., Jakovac, C.C. *et al.* (2021) Multidimensional tropical forest recovery. *Science* 374: 1370–1376. https://doi.org/10.1126/science.abh3629. See also: www.theguardian.com/environment/2021/dec/09/tropical-forests-can-regenerate-in-just-20-years-without-human-interference

Project Drawdown (2023) https://drawdown.org/drawdown-foundations

Rech, A.R., Ollerton, J., Dalsgaard, B. *et al.* (2021) Population-level plant pollination mode is influenced by Quaternary climate and pollinators. *Biotropica* 53: 632– 642.

Rech, A.R., Dalsgaard, B., Sandel, B. *et al.* (2016) The macroecology of animal versus wind pollination: ecological factors are more important than historical climate stability. *Plant Ecology and Diversity* 9: 253–262.

Remolina-Figueroa, D., Prieto-Torres, D.A., Dáttilo, W. *et al.* (2022) Together forever? Hummingbird–plant relationships in the face of climate warming. *Climatic Change* 175: 2. https://doi.org/10.1007/s10584-022-03447-3

Scerri, E.M.L., Roberts, P., Yoshi, M.S. and Malhi, Y. (2022) Tropical forests in the deep human past. *Philosophical Transactions of the Royal Society B* 377: 20200500. https://doi.org/10.1098/rstb.2020.0500. See also the other articles in the theme issue 'Tropical forests in the deep human past'. https://royalsocietypublishing.org/toc/rstb/2022/377/1849

Singh, Y. (2020) 'An ashram for the hummingbird': the Trinidad haven for world's tiniest bird. *The Guardian*. www.theguardian.com/environment/2020/jul/04/an-ashram-for-the-hummingbird-the-trindad-haven-for-worlds-tiniest-bird-aoe

Smith, A.C., Hurni, K., Fox, J. and Van Den Hoek, J. (2023) Community forest management led to rapid local forest gain in Nepal: a 29 year mixed methods retrospective case study. *Land Use Policy* 126: 106526. https://doi.org/10.1016/j.landusepol.2022.106526. See also: https://earthobservatory.nasa.gov/images/150937/how-nepal-regenerated-its-forests

UN Decade on Ecosystem Restoration (2023) www.decadeonrestoration.org
Walsh, S., Pender, R. and Gomes, N. (2022) Hawaiian endemic honeycreepers (Drepanidinae) are nectar robbers of the invasive banana poka (*Passiflora tarminiana*, Passifloraceae). *Journal of Pollination Ecology* 31: 8–15. https://doi.org/10.26786/1920-7603(2022)685

Acknowledgements

Birds and flowers represent ecological and evolutionary relationships. But this book is also about human relationships, because I could not have written it without the help and support of so many friends and colleagues – to all of whom I owe a debt of gratitude. Foremost amongst these are my wife Karin Blak and my good friend Nick Waser, both of whom took on the task of reading a first draft of each chapter as it was completed. Their comments and suggestions, and encouragement, greatly improved the final book.

So many other people have answered my queries and provided information (sometimes unsolicited) or have been instrumental in my published and unpublished research on birds and flowers, and more fundamentally my interest in understanding geographical and evolutionary patterns in specialised and generalised pollination systems. My sincere thanks to them all, and if I have missed anyone, my apologies: Jonas d'Abronzo, Ruben Alarcón, Felipe Amorim, Scott Armbruster, Peter Bernhardt, James Bullock, Cala Castellanos, Lars Chittka, Andrea Cocucci, Carolyn Coyle, Louise Cranmer, Bo Dalsgaard, K.D. Dijkstra, Mary Endress, Ted Fleming, Leonardo Galetto, Dennis Garber, Sjirk Geerts, Ana Martín González, Pablo Gorostiague, David Goyder, Dennis Hansen, Andrew Hingston, Richard Hobbs, Martin Hodson, Stěpán Janeček, Steve Johnson, Adam Karremans, Yannick Klomberg, Narayan Koju, Casper van der Kooi, Sabrina Aparecida Lopes, Francis Luma, Klaus Lunau, Chris Mackin, Mike Mann, Pietro Maruyama, Márcia Motta Maués, Nathan Muchhala, Ethan Newman, Stewart Nicol, Clive Nuttman, Kaz Ohashi, Jens Olesen, Ana Ortega Olivencia, Geoff Park, Anton Pauw, Kit Prendergast, Mary Price, Graham Pyke, Rob Raguso, Carsten Rahbek, John Reilly, Alejandro Rico-Guevara, André Rodrigo Rech, Zong-Xin Ren, Brody Sandel, Joey Santore, Ivan Sazima, Marlies Sazima, Henrik Schäfer, Alexander Schlatmann, Sheila Schueller, Alicia Sérsic, Jesper Sonne, Bill Sutherland,

Jens-Christian Svenning, Robert Tropek, Francisco J. Valtueña, Jeferson Vizentin-Bugoni, Stella Watts, Mikey Whitehead, the SURPASS2 team and many of the participants in Tropical Biology Association, Tenerife, Nepal and Brazilian field courses, and the SCAPE conferences. Thanks also to everyone who has commented on the posts on my blog (www.jeffollerton.co.uk/blog), where some of the writing in this book first appeared.

At the University of Northampton and Oxford Brookes University, Jo Alderton, Lynne Barnett, Andrew Hewitt, Janet Jackson, Andrew Lack, Duncan McCollin, Ian Moore, Lutfor Rahman, Kirsty Richards, Sam Tarrant and Stewart Thompson, amongst others, stimulated my wider interest in birds.

I'm grateful to the team at Pelagic Publishing, particularly Nigel Massen and David Hawkins, for their faith in the book, and to my good friend Stephen Valentine for his wonderful cover illustration. Particular thanks go to Hugh Brazier, whose editing skills and 'fine distinctions and nice judgements' have helped to make this a more readable, and accurate, book.

The generosity of funding bodies and institutions has given me the extraordinary privilege of carrying out field work, attending conferences and engaging in research projects across the world, and I am grateful for the support of (in no particular order): the University of Northampton, the University of New South Wales, Church's Shoes, the Royal Society, the Percey Sladen Fund of the Linnean Society, the British Ecological Society, the Leverhulme Trust, the Nuffield Foundation, the Royal Entomological Society, NERC, BBSRC, the Australian Research Council, the European Commission, FAPESP, Santander Universities, the Kunming Institute of Botany and the Chinese Academy of Sciences. Parts of this book were written during a month of teaching and field work in Kenya, organised by the Tropical Biology Association. I'm grateful to the staff and students on that course for their discussions and companionship as we explored the wonders of the savanna.

National and regional authorities gave permission to conduct field work in some parts of the world, and I owe thanks to them – as well as to those indigenous peoples whose land was taken by force in the colonial past.

Index

Abrahamczyk, Stefan 46–47, 137
Abutilon 156
Acacias (*Acacia* spp. and *Vachellia* spp.) 64, 80, 130–131, 195
Acanthus family (Acanthaceae) 47
Accentors (Prunellidae) 30, 33, 164
Adelaide, Australia 224
Africa 3, 17, 20, 23, 25, 28, 36, 39, 47–48, 70, 80, 86, 96, 104, 107–108, 112, 128, 130, 151, 191, 202, 209–212, 219, 223–224, 240, 246
African barbets (Lybiidae) 32
African Sacred Ibis 172
African Tulip Tree 198, 211
African violet family – see Gesneriads or african violet family (Gesneriaceae)
African warblers (Macrosphenidae) 30
Agave 242
Agriculture 5, 186–196, 207, 236
Agroforestry 193
AI – see ChatGPT and Artificial Intelligence (AI)
Aigner, Paul 62–63, 127
Ainsworth-Davis, J.R. (James Richard) 157
  Handbook of Flower Pollination (1906–1909) 157
Alarcón, Ruben 107
Alaska 39, 47, 105, 165
Ali, Sálim Moizuddin Abdul v, 90, 180–181
Aloe and red hot poker family (Asphodelaceae) 47, 58, 75, 93, 103, 136, Plate 2
Amani Nature Reserve, Tanzania 86, 152, 224
Amani Sunbird 152, 223–224
Amazon region, South America 104, 124, 193, 169, 230
Amber fossils 122
America, Central 34, 39, 104, 151, 199
America, North 3, 10, 16, 34, 37–38, 48, 66, 72, 104–105, 150, 155, 165, 172, 178, 183, 192, 230
America, South 3, 39, 58, 77–78, 108, 116, 123–125, 137–140, 149, 165, 167–170, 175, 184, 188–191, 201–205, 207, 211, 215–217, 229, 242–243
American Bird Conservancy 237
American Mistletoe 88
American Museum of Natural History, New York, USA 20
American Ornithological Society 176
Americas 11, 23, 33, 36, 39, 46–47, 50, 96, 167–170, 173, 202, 209, 212, 213, 216, 223, 234, 243
Amethyst Sunbird 1, 136
Amorim, Felipe 134, Plate 12
Amorim, Marsal 133
Amundsen, Trond 139
Amyema 90
Anagyris foetida 158–160, 164, Plate 15
Ancient Egyptians 172–173
Anderson, Bruce 25
Andes 25, 34, 44, 82, 85, 137–138, 140, 152–153, 230
Angiosperms
  Evolution of 8
  Diversity 43–44
Anna's Hummingbird 230
Anthreptes 23
Anthropocene 5, 212
Antillean Crested Hummingbird 111, 128

Aparecida Lopes, Sabrina 113–114
Apocynaceae – see Dogbane and milkweed family (Apocynaceae)
Araya-Salas, Marcelo 138–139
Argentina 50–51, 59, 114, 192, 208, Plate 8
Arnhem Land, Australia 181
Arteaga-Chávez, William 138
Asclepiads/Asclepiadoideae/ Asclepiadaceae – see Milkweeds (asclepiads)
Asclepias 51
Ashy Flowerpecker 180
Asia 28, 36, 38, 47, 90, 96, 104, 108, 125, 162, 165, 173, 185, 191, 201, 202, 218–219, 223, 225, 233, 243
Asities (Philepittidae) 31, 35–36, 88
Aslan, Clare 227
Atlantic Forest, Brazil 134, 203–205, Plate 12
Audubon, John James 97
Australasia 28, 36, 123, 224
Australia 1, 3, 6, 11, 16–17, 32, 36, 47, 60, 64, 76, 77, 89–90, 99–100, 104, 108, 118, 143, 181–182, 183–184, 190, 198, 202, 206, 209, 212, 213, 215, 217, 218, 220, 224, 225, 230–231, 234, 238, 239, 244
Australian Aboriginal rock art 181
Australian Museum, Sydney 11
Australian White Ibis 215
Avery, Mark 172
  A Message from Martha (2014) 172
Avian malaria 226, 236, 238
Axinaea 77–78, 152
Aztec culture 168–169

Bacteria (see also Nectar) 76, 117, 119–120, 211, 212, 222
Bagge, Niels Wessel 167
Baker, Herbert 73–74
Baker, Irene 73–74
Bald Eagle 22
Balsam family (Balsaminaceae) 47
Banana family (Musaceae) 53, 66, 190–191
Bananaquit 202, 239
Banksia 36, 47, 198, 220, 239, Plate 4, Plate 28
  Fossils 220
Barbados 176
Barlow, Roger 175
  A Brief Summe of Geographie (1500s) 175–176
Barnard, Phoebe 203
Barrowclough, George 20
Bascompte, Jordi 107
Bastiaan Meeuse 155, 157, 162
  The Story of Pollination (1961) 155
Bates, Henry Walter 124
  The Naturalist on the River Amazons (1863) 124
Bats as pollinators – see Mammals and flowers
Bauhinia 131
Beagle, H.M.S. 141, 145, 180
Beardtongues – see Penstemons
Bee Hummingbird 11–12, 149–150
Bees (Hymenoptera) 8, 12, 13–14, 40, 43, 50, 51–54, 62, 69, 73, 84, 90, 105, 110, 118, 125, 130–131, 134, 137, 148, 149, 152, 170, 178–179, 184, 185–188, 194, 205, 207, 212, 227, 229
  Bumblebees 58, 63, 83, 119, 133, 152–153, 175, 196, 216, 231–232
  Carpenter bees 86
  Honey bees 39, 80, 83, 171, 173, 185, 187, 224
  Honey collecting 171
  Orchid bees 93
  Sensitivity to red 95–96
  Solitary bees 83, 119
Begonias 87
Bellflowers 149
Bent, Arthur Cleveland 104
  Life Histories of North American Wood Warblers (1953) 104
Berrypeckers and longbills (Melanocharitidae) 31, 35, 153
Betts, Matthew 115
Bicácaro – see Canary Island Bellflower

Bills/beaks adapted for nectar-feeding 11, 34, 36, 37, 83–84, 88, 89–90, 126–129, 135–140, 149, 176, 180, 203, 209, 214, 227
Bimodal pollination systems – see Pollination systems
Biodiversity crisis 220–232
'sixth mass extinction' 220–221
Biodiversity Research Collections, University of Connecticut, USA 122
Biogeographic and geographic patterns 4–5, 28, 44, 51–52, 96–97, 112, 125, 128, 145, 164–166, 179, 228, 230, 242
Bird behaviour
Aggression 10, 38, 85–86, 90, 111, 204–205, 218, 232, Plate 28
Brain size and organisation 94–95
Medicinal use of flowers 3, 35
Migration 39, 42, 98, 101, 103–105, 153, 165, 179, 201, 231, 235, 245
Traplining 94, 119, 127
Bird diversity 7, 19–20, 43
Bird feeders 24, 38, 98, 102, 118, 132, 138, 203–206, 230, 235
Bird of Paradise (plant) iv, 24, 66–68, 157
Bird physiology 92, 94, 129, 131–132, 135
Bird pollination
Advantages of 25, 54–56, 73
Frequency of 45–46
Origin of 9–10, 12–15, 88
Bird senses 4, 92–100
Magnetic field detection 93–105
Smell 97–99
Taste 99–100
Vision 4, 92–97, 99
Bird's-foot-trefoils (*Lotus* spp.) 87, 146, 148, 228
BirdLife International 221
Birds Evolution of 7, 10–13
Birds of paradise (Paradisaeidae) 22, 138
Birds of the World 5, 19–20, 32, 35, 39, 104, 223
Bismarck Black Myzomela 217
Black Currawong 32
Black Mamo 128
Black Sunbird 41
Black-capped Bulbul 136
Black-chinned Hummingbird 105
Blak, Karin 13, 80, 109–110, 167, 241, 245
The Essential Companion to Talking Therapy (2021) 109
Blue Larkspur 133
Blue-Faced Honeyeater 190
Blue-headed Hummingbird 128
Blue-mantled Thornbill Plate 3
Bluethroat 139
Bohemian Waxwing 22
Boles, Walter 11
Bolivia 208
Bottlebrush (*Callistemon* sp.) 1
Bowerbirds (Ptilonorhynchidae) 3, 138
Boyd, Rob 229
Brachychiton 211
Brahminy Starling 35
Brazil 17, 46, 63, 78–79, 107, 112–113, 124, 130–134, 156, 176–177, 179, 183, 185, 188–190, 193, 198, 201–205, 213, 235, 238
Bredenkamp, Henry 103
British Ecological Society 207
British Museum, London 172
'Nebamun hunting in the marshes' 172
Broad-tailed Hummingbird 105, 106, 206
Brooks, Dan 124–125
Brown Rat 190
Brown Tree Snake 196
Buff-tailed Coronet Plate 1
Buglosses (*Echium*) 146, Plate 14b
Bulbuls and allies (Pycnonotidae) 29
Busch, Jeremiah 103
Bushshrikes and allies (Malaconotidae) 30
Bushwillow (*Combretum* sp.) 78–79
Butea 211

Butterflies (Lepidoptera) 8, 54, 63, 71, 80, 83, 110, 135, 191
African Monarch 172
Heliconius 77
Monarch 51

Cabot, Sebastian 175
Cáceres, Brazil 201
Cactus family (Cactaceae) 40, 49–51, 59, 77, 85, 192–193, Plates 8 & 10
Calder III, William A. 39
California 6, 15–17, 119, 184, 187
Camera traps 65, 71, 111, 134
Cameroon 49, 111–112
Campinas, Brazil 197–198
Canada 165
Canary Island Bellflower or Bicácaro 148, 192
Canary Island bird flowers 18, 72–73, 144–149, 160, 192, 212, 227–228, Plate 14
Canary Island Chiffchaff 18–19, 147
Canary Island Foxglove 18, 72, 148, 212
Canary Islands 3, 5, 17, 18–19, 144–149, 154, 160, 192, 209
Cape Floral Region, South Africa 122–123, 137, 203, 205, 235
Cape Glossy Starling 136
Cape Sugarbird 122, 184, Plate 24
Cape Town, South Africa 203, 205, 235
Cape Weaver iv, 66, 157
Carboniferous 92
Cardinals and allies (Cardinalidae) 30
Cardoso, João Custódio Fernandes 190
Caribbean 39, 93, 111, 143, 149–151, 177–179, 223, 235, 242
Carlos Herrera 119
Carolina, USA 177–178
Case, Ted 121
Community Ecology (1986) 121–122
Castañeda-Zárate, Miguel 136
Castellanos, Maria Clara 'Cala' 52–53, 216, 241
Catesby, Mark 177–179
The Natural History of Carolina, Florida and the Bahama Islands (1729–1747) 177–179
Cetti's Warbler 230
Chalcomitra 23
Chapman, Hazel 112–113
Charles University, Czech Republic 111
ChatGPT and Artificial Intelligence (AI) 22, 162–163, 243
Chatham Island Bellbird 224
Chen, Zhe 96–97
Cherry Plum 14, 161
Chile 44–45
China 3, 38, 104, 165, 191–192, 197, 199–201, 209, 219, 240
Chinese Academy of Sciences 200
Chinese art 173–174, Plates 18 & 19
Chittka, Lars 95, 145, 147
Chmel, Kryštof 131–132
Christidis, Leslie 217
Cisticolas and allies (Cisticolidae) 29
Citizen science and conservation 42, 28, 105, 234, 236
Journey North project 105
Songbirds as Pollinators (SaP) project 42
Classen-Bockhoff, Regine 68
Clements, Frederic Edward 106
Experimental Pollination: an Outline of the Ecology of Flowers and Insects (1923) 106
Climate change 5, 114, 123, 165, 187, 192, 210, 220, 221, 231, 234, 236, 240–241, 243–244
Coal Tit 153
Cockatoos (Cacatuidae) 32, 36
Cocucci, Andrea 135
Codependency 109–110
Coetzee, Anina 203
Co-evolution 126–128, 132, 137–140
Arms races between species 128, 138–139
Coffee Berry Borer 196
Colombia 34, 152190, 216, 230, 238
Colorado State University, USA 42

Colwell, Robert 121–122
Common Cactus-Finch 77, 143
Common Chaffinch 34
Common Chiffchaff 14, 159, 200, Plate 15
Common Foxglove 216–217
Common Guava 143
Common Myna 190.218
Common Primrose 161
Common Scale-backed Antbird 124
Common Starling 217
Common Sunbird-Asity 36
Common Wood-Pigeon 14
Common Yellowthroat 67
Compositae (see also Daisy family (Asteraceae)) 44, 104
Conservation 8, 89, 132, 149, 183, 223–226, 228, 229, 233–246
  Of Hawaiian honeycreepers 29, 34, 128, 226, 236–238
Coogee Bay, Australia 234, Plate 4
Cook, Captain James 226
Coombs, Gareth 157
Copenhagen, Denmark 167
Copper-throated Sunbird 191
Coral trees (*Erythrina* spp.) 19, 73, 202
Corfield, Jeremy 97
Corlett, Richard 47, 219
Cornell Lab of Ornithology 5, 19, 32, 199
Cornell University, USA 98
Corvus 33
Cosnett, Emmeline 58
Costa Rica 46, 100, 104, 115, 121, 127, 205, 216, 238–239
Costa's Hummingbird 40, 49, 105
Cotton, Peter 104
COVID-19 pandemic 50, 80, 213, 245
Coyle, Carolyn 42
Coyote Tobacco 210
Crakes and rails (Rallidae) 12
Cranmer, Louise 145, 147–148
Crested Auklet 139
Crested Berrypecker 35, 81
Cretaceous 8, 38, 122
Crimson Sunbird 192
Crop and wild harvested plants
  Acca sellowiana – see Feijoa
  Bacuri 193–194
  Bananas 53, 190–191
  Cacao 187
  Coffee 80, 86, 104, 187, 196, 205
  Date palm pollination 173
  Durian 191
  Feijoa 188–190, 194, 196, 218
  Feijoa sellowiana – see Feijoa
  Guaba 188, 239
  Indigenous crops 192–193
  Kapok 195
  Loquat 192
  New Zealand flax 58, 195
  Non-food crops 188, 194–195
  Papayas 191
  Pineapples 191
  Pomegranate 174, 185–186, Plate 19
  Saguaro fruit harvesting 193
  Silk-cotton 195
  Wild or Donne' sali Chilli 196
Crow Honeyeater 224
Crows, jays and magpies (Corvidae) 29, 33, 197
Cruden, Robert 152
Cuba 11, 37–38, 149–151
Cuban Emerald hummingbird 38
Cuban Green Woodpecker 37–38, Plate 5
Cuckoos (Cuculidae, Cuculiformes) 32, 40, 143
Cultural influences of birds & flowers 2, 5, 18, 153, 162, 167–174, 182–184, 242–243
Curtis's Botanical Magazine 212
Cyanomitra 23
Cypripedium 164
Czech Republic 111

d'Abronzo, Jonas 203–204, 235
Daisy family (Asteraceae) 44, 47, 69
Dalsgaard, Bo 37–38, 107, 111, 128, 150–151, 167, 244
Darjeeling Woodpecker 38
Dark-billed Cuckoo 143

Darwin, Charles 8, 34, 97, 117, 139, 141–142, 143, 145, 154, 156–157, 161, 180, 183–184
The Effects of Cross and Self Fertilisation in the Vegetable Kingdom (1876) 157
'Survival of the fittest' 117
Zoology of the Voyage of H.M.S. Beagle (1841) 180–181
Darwin, Erasmus 161–164
The Botanic Garden. A poem in two parts (1791) 162
Darwin's (or Galápagos) finches 29, 34, 103, 141–144
Daturas 138, 139
Davidar, Priya 90–91
de Alcantara Viana, João Vitor 124
de Léry, Jean 179
Histoire d'un voyage fait en la Terre du Bresil, autrement dite Amerique (1578) 179
del Coro Arizmendi, María 205
Dellinger, Agnes 77–78, 152
Delonix 211–212
Denmark 3, 184, 213
Devonian 7
Diamond, Jared 121, 217
Community Ecology (1986) 121–122
Dicaeum 35
Diller, Carolina 136
Dinosaurs 7–9, 117, 146
Dirzo, Rodolfo 238
Diversity and Distributions 213
Djulirri, Australia 181
DNA, RNA and molecular systematics 20, 22–23, 28, 34, 39, 41, 51–53, 101–102, 115, 184, 200, 220, 225
Dogbane and milkweed family (Apocynaceae) 44, 51, 63, 69, 70
Doherty, James (Tahae) 171
Dominica 127–128
Dragonflies (Odonata) 125
Drongos (Dicruridae) 30
du Plessis, Monique 205
Duchenne, François 114
Dupont, Yoko 147, 148
Duranta 82–85, 133
Dusky Sunbird 209
Dutch East India Company 179

East Usambara Mountains, Tanzania 86, 224
Eastern Arc Mountains, Africa 151–152
Eastern Olive Sunbird 86
Eastern Spinebill 183
eBird 199
Echinopsis 51, 59, Plate 8
Ecology (journal) 16–17
Ecosystem disservices 195
Ecosystem restoration – see Habitat and ecosystem restoration
Ecosystem services 184, 195–196, 198
Ecuador 24, 54, 98, 223, 230
Eda, Masaki 168
eDNA 243
Edward Grey Institute of Field Ornithology, UK 142
Egyptian Vulture 172
Elegant Sunbird 223, 225
Endemic species/Endemism 32, 35–36, 41, 44, 47, 51, 143, 148, 149, 151, 153, 171, 196, 219, 223–228, 230, 236, 239, 244
Enemy Release Hypothesis (ERH) 212–213
Eocene 12
Erythrina (see also Coral Tree) 211, Plate 7
Ethiopia 240
Eucalyptus 16–17, 36, 188, 195
Eurasia 165–166
Eurasian Blackbird 190, 218
Eurasian Blackcap 35, 159
Eurasian Blue Tit 14, 161, Plate 16
Eurasian Collared-Dove 206, 230
Europe/Europeans 3, 5, 11–12, 23, 35, 41, 88, 104, 145, 146, 154, 155–166, 171, 174, 179–180, 182, 185, 189, 197, 200, 206, 208, 209–210, 215–216, 225–226, 230, 242, 246
European Mistletoe 88, 196
Eurotrochilus 11

Evolution of flight 7, 64
Evolution of insect pollination 7–8
Ewaso Ny'iro River, Kenya 245
Extinction and extinct species 8–10, 15, 19–20, 27, 34, 39, 43, 128, 147, 165, 168, 178, 220–223, 224, 225–226, 236
Extirpation 196, 221–228
Extrafloral nectaries 64

Facebook 241
Fairy-bluebirds (Irenidae) 31
Falcons (Falconiformes) 22, 24
Fan, Tai-His 129
Fantails (Rhipiduridae) 30
Farlow, James 8–9
Farmer, Chris 237
Feral Pigeon 199
Ferguson, Theo and Gloria 235
Figs (*Ficus*) 135, 198
Figwort family (Scrophulariaceae) 53, 160
Finches, euphonias and allies (Fringillidae) 3, 17, 28, 29, 34, 88, 124
Fire-breasted Flowerpecker 219
Firecrowns (*Sephanoides* spp.) 39
Fire-tailed Myzornis 35, 129, 174
Fire-tailed Sunbird 153
Fitzsimons, James 33
Flame of the Forest 211–212
Flies (Diptera) 13–14, 43, 97, 125, 130, 131, 135, 187, 195,
  Hoverflies/syrphids 77, 83, 187, 229
  Long-tongued flies 54
Florivory 13–14, 80–82, 87–88, 171
Flowering plants – see Angiosperms
Flowerpeckers (Dicaeidae) 14–15, 24, 30, 35, 88, 90–91, 180, 219
Flowerpiercers (*Diglossa* spp.) 34, 78, 83–84
Flowers
  Colouration 62–63, 71, 94–96, 131, 173
  Evolution of 9, 16, 62–65
  Female function 67, 84
  Flower lifespan 121, 148
  Flower scent 50, 59, 62, 97–99
  Male function of 67, 84, 88
  Red flowers 37, 49, 76, 93–97, 185, 211
  Rewardless 9–10, 74, 81–82, 86
  Scentless flowers 58, 97–99, 131, 159
  Self incompatibility 85, 190
  Self pollination 54, 209, 227
  Sexual systems – dioecy, etc. 67, 86
  Spectrophotometer 57
  Zygomorphy 90
Fontúrbel, Francisco 44–45
Food and Agriculture Organization of the United Nations (FAO) 194
Food system resilience 186–187
Fork-tailed Sunbird 219
Fossil birds 10–13, 15, 20, 39, 138, 147, 165, 220
Fountain, Jade 218
Foxgloves (*Digitalis*) 18, 72–73, 148, 212, 215–217, 146, Plate 14a
France 11
Frick, Karen 239
Frogs as (possible) pollinators 149
Frugivory and fruit-eating 10, 12, 15, 34, 35, 36, 37, 40, 83, 89, 91, 102, 103, 144, 172, 180, 189, 198, 219
Fruit-eating – see Frugivory and fruit-eating
Fuchsia 57, 58, 156, 211
Fukami, Tadashi 119–120
Fungi 118, 120, 122, 151, 212–213, 222

Gabriela Boaventura, Maria 87
Galápagos Dove 143
Galápagos finches – see Darwin's finches
Galápagos Flycatcher 143
Galápagos Islands 5, 77, 103, 141–144, 145, 184
Gamba, Diana 54–55
Garcinia family (Clusiaceae) 193
Garden Warbler 162
Gardens 1, 5, 13–14, 38, 51, 57–58, 66, 73, 125, 156, 161–162, 188, 190, 193, 199–200, 203–206, 209, 215, 218–219, 224, 226, 228, 235, 245

Gavini, Sabrina 114
GBIF – see Global Biodiversity Information Facility
Geerts, Sjirk 25, 137
Geiger Tree 37, Plate 5
Generalist flowers and pollination systems 16, 75, 109, 112–114, 126, 130, 140, 210
Generalist passerines 47, 148, 149, 157, 164, 173, 192, 203
Genes, Luísa 238
Geographic mosaic theory 128
Geographic patterns – see Biogeographic and geographic patterns
Germany 12, 39
Gesneriads or african violet family (Gesneriaceae) 47, 55, 120, 140
Ghana 191
Giant Hummingbird 39, 123, 181
Gibraltar 162, 202
Gila Woodpecker 38, 40, 49
Glittering-bellied Emerald Plate 10
Glittering-throated Emerald 201
Global Biodiversity Information Facility (GBIF) 229–230
Global Invasive Species Database 210
Godsoe, William 112–113
Goldcrest 24, 180
Golden Camellia 192
Golden Wattle 64
Golden-winged Sunbird 128
Goliath Tarantula 123–124
Gomes, Noah 236
Gorostiague, Pablo 50, 59, 192, 241, Plates 8 & 10
Gottsberger, Gerhard 45–46, 55
Graham, Catherine 114
Grain for Green project 240
Gran Canaria Blue Chaffinch 145
Granivory 34, 142, 196
Grant, Karen 53–54
  Flower Pollination in the Phlox Family (1965) 53, 54
Grant, Rosemary 77
Grant, Verne 53–54
  Flower Pollination in the Phlox Family (1965) 53, 54
Gray-hooded Sierra Finch 59, Plate 8
Great Green Wall project 240
Green Woodhoopoe Plate 6
Green-backed Firecrown 165
Green-backed Honeyeater 103
Greengage 13–14
Green-throated Carib 128
Grenada 128
Grevillea 47, 104, Plate 25
Grew, Nehemiah 176, 180
Grime, Phil 207
Guaba 188, 239
Guam, Mariana Islands 196
Guanches 192
Guans, chachalacas and curassows (Cracidae) 32
Gulls and shorebirds (Charadriiformes) 12, 22, 40, 31, 78
Guyana 17, 116, 125
Gymnosperms 7–8

Habitat and ecosystem restoration 234–235, 237–241
Habitats and plant communities 15, 35, 46, 89, 101, 138, 141–142, 157, 227, 229
  Afromontane 113
  Atlantic Forest 134, 203–205, Plate 12
  Caatinga 46
  Campo Rupestre 63, 113–114, 133
  Cerrado 46, 55, 130
  Cloud and humid forest 54, 82, 152
  Disturbance and loss 82, 114–115, 218, 219–221, 223, 226
  Ecotone 116
  Forest 7, 12, 15, 23, 45, 46, 82, 89, 112, 172, 218, 236, 242
  Fynbos 205
  Gallery forest 55
  Grassland 15, 24, 45, 63, 89, 107, 113, 116, 133, 234
  Harvesting from 194–195
  Laurel forest 146

Páramo 152
Rainforest 16, 35, 46, 86, 104, 112–113, 115, 116, 204, 206–207, 224, 242
Restoration of – see Habitat and ecosystem restoration
Rhododendron forest 153
Savanna 80, 64, 116, 130, 245
Stressful – for a penguin 207
'Tertiary relicts' 146, 154, 158, 164
Urban and suburban 197–207
Hachuy-Filho, Leandro 124
Hadley, Adam 115
Harper, John 207
Hawaii 11, 34, 128, 210, 219, 225–227, 236–238
Hawaii Akepa 219
Hawaii Amakihi 236–237, 238
Hawaiian honeycreepers (Fringillidae, subfamily Carduelinae) 29, 34, 128, 226, 236–238
Hawaiian honeyeaters (Mohoidae) 31, 225–226, 236
Hawaiian lobelioids (or lobeliads) 226–227
Hawks, eagles, vultures, and kites (Accipitriformes) 22
Vultures and condors 97
Heard, Stephen 163
Heather and rhododendron family (Ericaceae) 32, 35, 205
Heliconias (*Heliconia* spp.) 115, 124, 126–128, 227
Henry VIII 175
Herbivory 8, 10, 13, 15, 37, 64, 90115, 117, 144, 212
Herring Gull 22
Hilpman, Evan 103
Himalayas 25, 153, 174
Hingston, Andrew 60
Hispaniola 151, 174, 179
Hispaniolan tanagers (Phaenicophilidae) 31
Historical discovery of bird pollination 5, 167–184
Hobbs, Richard 183
Holly 196
Homo 22, 221–222
Honeyeaters (Meliphagidae) 1, 4, 11, 17, 19, 23, 29, 32, 33, 35, 39, 47, 64, 71, 77, 88, 104, 108, 112, 123, 124, 125, 129, 152, 171, 181–182, 183, 184, 190, 209, 219, 220, 234, Plates 25, 28
Arthropods, pollen and fruit in their diets 77, 102–103
Conservation of 224
Early European accounts of 181–182
Fossils 11, 220
Indigenous knowledge of 181
Mistletoe pollination 89–90
Range expansion and migration in 153, 217–218, 230–231
Honeyguides (Indicatoridae) 32
Hong Kong, China 6, 197, 219
House plants 51, 70
House Sparrow 87, 199, 201, 217
Hovering 11, 16, 24–26, 66–67, 95, 126, 209
Huang, Shuang-Quan 192
Hughes, Alice 199
Humboldt Brewing Company, California Plate 23
Hummingbirds (Trochilidae) 2, 3, 14–17, 18, 27–28, 31, 46–47, 49–50, 58, 66, 68, 70–73, 88, 106, 107–108, 111–115, 116, 118–125, 126–134, 136–140, 147, 149–152, 155–156, 163, 165, 181, 182–184, 185–186, 189, 199, 208–211, 227, 239, Plates 1, 3, 10, 17, 20, 21, 23, 24, 26
Arthropods in their diets 75, 100–103
As ineffective pollinators 83–85
Conservation of 223, 228–230, 235, 239, 244–245
Crop pollination 185, 192–195
Early European accounts of 174–179
Fossils 11–12, 23, 39, 165
Hovering in 24–26

In gardens and urban areas 201–206, 235
Indigenous knowledge of 167–170, 242–243, Plate 17
Migration in 39, 101, 103–105, 179, 201
Pollen digestion 77
Pollinating non-native plants 212–217
Pollination systems/syndromes 53–55, 62–63, 93–96
Sense of smell 98–99
Sicklebill 126–128
Taxonomy 14, 39–40
Hunter, Fiona 139
Huxley, Elspeth 211
Huxley, Sir Julian 142
Huxley, Thomas 142

Icterus 33
Iiwi 128, 227, 236–237
Ilha Solteira, Brazil 202
Imperial (or Crown) Fritillary 162, 163, 173
India 85, 90, 143, 180, 185–186, 212, 240
Indian paintbrushes (*Castilleja* spp.) 103–104, Plate 11
Indian Peafowl 138
Indian Pied Starling (or Myna) 19
Indigenous and Local Knowledge about Pollination and Pollinators Associated with Food Production (2015) 170
Indigenous peoples and knowledge 5, 153–154, 167–184, 192–193, 225–226, 235, 242–243
Indomalayan region 47
Indonesia 41, 155, 179, 191, 200
Ingcungcu Sunbird Restoration Project 235
Inouye, David 206
Insectivory 10, 13–15, 24, 37, 88, 150
Insects as pollinators 3, 7–9, 13–14, 43, 47, 58, 64, 75, 77
Interaction networks 4, 86–87, 105, 106–115, 227
'Bipartite' graphs 106
Modularity 110, Plate 13b
Nestedness 109–110, Plate 13a
Rewiring 115
Interaction release 144, 148–149, 150, 202
Interaction types 117–118
Commensalism 10, 117, 122, 212
Mutualistic interactions/mutualism 9, 10, 40, 81–82, 91, 117, 180, 212–213
Parasitism 10, 34, 81, 88–89, 115, 117, 134, 158, 212, 222
Phoresis 122–123
Predation 10, 14, 123–125, 172, 212
Interaction Web Database 106
Intergovernmental Platform on Biodiversity and Ecosystem Services 170
International Botanical Congress 200
International Union for Conservation of Nature (IUCN) 19, 223, 228
Invasive alien species (IAS) 5, 25, 143, 172, 196, 208–219, 220, 223, 225–226, 236–237, 239, 241, 242
IOC *World Bird List* 20, 26
IPBES – see Intergovernmental Platform on Biodiversity and Ecosystem Services (IPBES) 170
Iran 173
Ireland 58
Iris family (Iridaceae) 25, 47
Island Canary 34
Islands 4–5, 39, 44, 70, 111, 127–128, 140, 141–154, 196, 202, 217, 223–228
Extinction rates on 223, 224, 225–226, 236
Israel 206, 209, 210
Ivory-billed Woodpecker 178
Izhaki, Ido 210

Janeček, Štěpán 25, 111
Janzen, Dan 222, 226
Javan Sunbird 191
Jayanthi, P.D. Kamala 186
Johnson, Steve 60, 69, 74–75, 136

Jones, Ian 139
Jones, Steve 183–184
Jønsson, Knud 115
Jordano, Pedro 107
Journal of Ornithology 205
Journal of Pollination Ecology 209, 236
Juan Fernandez Firecrown 223
Juan Fernández Islands 39, 223
Jurassic 6, 7–8
Jurassic Park 222

Kaiser-Bunbury, Christopher 239
Kalamazoo, USA 183
Kallarackal, Jose 189
Kauai Oo 225
Kay, Quentin 161
Keighery, Greg 47
Kelly, Dave 89–90
Kenya 1, 17, 41, 80–81, 130, 151, 245–246
Kessler, Michael 46–47
Kestrel 24
Keystone species 89
King Prawn Curry mnemonic 21
Kinglets (Regulidae) 31, 180
Klein, Alexandra-Maria 187
Kniphofias (*Kniphofia* spp.) 58, 93
Knudsen, Jette 98–99
Knuth, Paul 157
Handbook of Flower Pollination (1906–1909) 157
Koju, Narayan 174
Krauss, Siegfried 239
Kudom, Andreas 191
Kunming Institute of Botany, China 174
Kwapong, Peter 191
KwaZulu-Natal, South Africa 107

La Gomera, Canary Islands 147
La Selva Biological Station, Costa Rica 46, 121, 127
Lack, Andrew 6, 142–143
Lack, David 142, 150, 151, 154
Ladley, Jenny 89–90
Lake District of England 242
Landfowl, quails, megapodes, etc. (Galliformes) 21–22, 32, 40, 94
Langtang National Park, Nepal 153
Las Cruces Biological Station, Costa Rica 115
Latham, John 181
Index Ornithologicus (1790) 181
Laughing Dove 124
Laughingthrushes and allies (Leiothrichidae) 29
Lavenders (*Lavandula* spp.) 58
Leaf warblers (Phylloscopidae) 14, 25, 30, 103, 153
Leafbirds (Chloropseidae). 31, 35
Least effective pollinator principle (LEPP) 62–63, 127
Least Seedsnipe 78, 135
Lebanon 165
Legumes – see Pea and bean family or legumes (Fabaceae)
Leimberger, Kara 115
Lek 139
Lepcha people 174
Lesser Antilles 127–128
Lesser Honeyguide 39
Leven, Michael 219
Lewis, Gwilym 212
LIDAR 243
Light-vented Bulbul 192
Lindell, Catherine 238
Linnaeus, Carl 159
Little Egret 230
Little Spiderhunter 191
Lizards as pollinators 3, 9, 43, 144, 148
Long Island, New Guinea 217
Long, Frances Louise 106
Experimental Pollination: an Outline of the Ecology of Flowers and Insects (1923) 106
Long-bearded Melidectes 152
Long-tailed Hermit 100
Long-tailed tits (Aegithalidae) 30, 164
Lord Tennyson, Alfred 117, 123
'Nature, red in tooth and claw'. 117
Lories and lorikeets (subfamily Loriinae) 17, 36, 77, 220
Lorikeets – see Lories and lorikeets (subfamily Loriinae)

Louisiana Museum of Modern Art, Denmark 167, 169
Loveridge's Sunbird 152
Low, Tim 99–100
Where Song Began: Australia's Birds and How They Changed the World (2016) 99–100
Lunau, Klaus 93–94
Luzon, Philippines 152

Mackin, Chris 216, 241
Macquarie University, Australia 16
Madagascar 35, 198, 211–212
Magner, Evin 76
Malachite Sunbird 137, 209, Plate 2
Malaysia 123
Mallow family (Malvaceae) 93, 148
Mambilla Plateau, Nigeria 112–113
Mammals and flowers 19, 87, 134
Bats 3, 19, 24, 40, 43, 50, 54, 134, 137, 191, 194, 229
Coatis 134
Monkeys 19, 80, 87
Opossums 134
Rodents 19, 87, 134, 190
Tayras 134
Mangroves 123
Mann, Michael 233–234
Hockey stick graph 233
The New Climate War (2021) 234
Mantids – see Praying mantis
Māori people 171–172
Marcgrave, Georg 179
Mariana Islands, Micronesia 196
Mariqua Sunbird 130, Plate 27
Marlies Sazima 78–79
Maruyama, Pietro 130–131, 133, 201, 213
Masked Flowerpiercer 83
Massufaro Giffu, Murilo 124
Matallana-Puerto, Carlos 190
Mato Grosso, Brazil 201
Maui Forest Bird Recovery Project, Hawaii 237
Maung people 181
Mauritius 149, 238–239
Mauritius Gray White-eye 239
Mayfield, Margie 63
Mayr, Gerald 11–12
McDonald, Paul 218
McQuillan, Peter 60
Mediterranean 3, 158, 165, 210
Medium Ground-Finch 77, 143
Melbourne, Australia 231
Memmott, Jane 214
Merian, Maria Sibylla 123–124
Metamorphosis Insectorum Surinamensium (1705) 123
Messel Shale 12–13
Mexico 101, 104, 169, 119, 205, 208, 244
Micheneau, Claire 71
Microbial Ecology (journal) 122
Middle East 173
Milkweeds (asclepiads) 44, 51, 63, 69–72, 130, 160
Mimicry by plants and animals 86, 90, 91, 135, 164
Mimosaceae (see also Pea and Bean family (Fabaceae)) 104
Minas Gerais, Brazil 133
Mint and sage family (Lamiaceae) 68, 90, 227
Miocene 11, 140, 220
Missed Mutualist Hypothesis (MMH) 213
Mistletoes 88–91, 180, 196
Mites (Arachnida) 117, 121–123
Mittelbach, Moritz 119
Mnisi, Bongani 235
Mockingbirds and thrashers (Mimidae) 30, 135, 143, 189
Molecular systematics – see DNA, RNA and molecular systematics
Moles, Angela 213
Monkeyflower (*Mimulus* sp.) 119
Monkeys – see Mammals and flowers
Montgomerie, Bob 176
Monty Python and the Holy Grail 87
Moorland Chat 41, Plate 7
Moran, Alison 101–102
Moran, Jonathan 101–102
Most effective pollinator principle (MEPP) 62–63

Moths (Lepidoptera) 8, 40, 43, 51, 52, 54, 59, 80, 83, 96, 195
Hawkmoths 63, 155
Yucca moths 135
Mount Cameroon 111–112
Mount Kenya 1, 41,
Mount Kenya lobelias 41, Plate 6
Mount Teide Bugloss Plate 14b
Mountain Chat – see Moorland Chat
Mountain Sunbird 152
Mountains 4, 18, 86, 140, 141–154, 168, 224, 236, 242
Mousebirds (Coliidae, Coliiformes) 31, 40, 80, 81
Moussonia 120
Mpala Research Centre, Kenya 130, 245
Muchhala, Nathan 54–55
Müller, Fritz 156
Müller, Hermann 155–157
Die Befruchtung der Blumen durch Insekten (1873) 156
The Fertilisation of Flowers (1883) 156
Mundi, Onella 49
Muschett, Giselle 44–45
Music and musicians 18, 64, 120, 182–183
Gibson Guitar Corporation 182–183, Plate 20
Guitars 182–183, 184, Plates 20, 21
Jones, Brian 183
King, B.B. 182
Lane, Frankie 182
Led Zeppelin 64
'Oxford Scene' 120
Page, Jimmy 182
Paul Reed Smith (PRS) Guitars 183, Plate 21
Radiohead 120, 183
Richards, Keith 183
Ride 120
Rolling Stones 183
Snider, Hartford 183
Songs and albums 182
Stevens, Cat 182
Supergrass 120
Swervedriver 120
The Nature of Music blog 183
Yorke, Thom 183
Mute Swan 22
Mutualism 9, 10, 40, 81–82, 91, 117, 180, 212–213
'Biological barter' 67
Cheating in 9–10
Double 40, 91, 180
Myanmar 104, 143
Myrtle and eucalyptus family (Myrtaceae) 188

Nahua people 177
Nakula Natural Area Reserve, Hawaii 237
Natural History Museum at Tring, UK 141
Nature (journal) 71–72, 161
Nazca culture 167–170, Plate 17
Nectar characteristics and analysis 8–9, 57–58, 72–76, 83, 100, 118–120, 147, 210
Alcohol in 118
Anabasine 210–211
Bacteria in 76, 117, 119–120
Coloured 76, Plate 9
Fermentation of 118–120
Nicotine 210–211
Sucrose:hexose (S:H) ratio 73–74
Sugar refractometer 57, 72
Sugars 57–58, 72–76, 83, 99, 118, 120, 129
Nectar robbing 34, 58, 83–85, 86, 90, 98, 185–186, 202, 237
Nectarivory 15, 24–25, 27, 33, 35, 73–76, 100, 102, 103, 132, 134–136, 149, 226
Neogene period 146
Neognaths 21
Neotropics 2, 38, 55, 93, 100, 229
Nepal 3, 17, 38, 153, 174, 236
Networks – see Interaction networks
New Caledonia 224
New England, USA 178
New Guinea 35, 41, 152, 153–154, 217
New Holland Honeyeater 77, 181

New South Wales, Australia 104, 213, 234
New World and African parrots (Psittacidae) 32, 36
New World Sparrows (Passerellidae) 29, 33, 135
New World warblers (Parulidae) 29
New Zealand 3, 6, 11, 16, 36, 58, 89–90, 156, 171–172, 190, 195, 209, 218, 225, 232
New Zealand Bellbird 89, 156
New Zealand Kaka 36
New Zealand parrots (Strigopidae) 32, 36
New Zealand Pigeon or *Kererū* 171–172
New Zealand wrens (Acanthisittidae) 27, 31
Newstead, Australia 230–231
Nicol, Stewart 32
Nicolson, Susan 74–75
Nigeria 112–113, 191
Nile Valley 172–173
Nile Valley Sunbird 172–173
Niven, Jeremy 95
Noisy Miner 206, 218, 232, 234, Plate 28
Norfolk Island 225
Norman, Janette 217
North Africa 146–147
Northampton, UK 13, 58, 103, 145, 188, 219
Nsor, Charles 112–113
Nuytsia 211
Nyffeler, Martin 125, 195

O'Hanlon, Redmond 123–124
Ocotillo 105
Ocotillo 105
Oikos 158
Old World buntings (Emberizidae) 30
Old World flycatchers (Muscicapidae) 29, 41, Plate 6
Old World orioles (Oriolidae) 30, 33–34, 164
Old World parrots (Psittaculidae) 28, 32, 36, 219–220
Old World sparrows (Passeridae) 22, 30, 87, 157, 199, 202, 217
Olesen, Jens 147, 149
Oligocene 11
Olive-backed Sunbird 41
Omnivory 33, 103
Optimal foraging theory 83, 95
Orange-breasted Sunbird 123, 183
Orchid family (Orchidaceae) 65, 69–71, 74, 135, 164, Plate 26
  Arpophyllum giganteum Plate 26
  Cypripedium 165
  Moth orchids (*Phalaenopsis*) 70
Øresund, between Denmark and Sweden 167
Ornithophily and ornithophilous flowers 48–49, 55, 61–62, 63, 75, 93, 130–132, 157
Ortega-Baes, Pablo 50, 59
Ortega-Olivencia, Ana 158–160
Ortiz-Crespo, Fernando 177
Ovenbirds and woodcreepers (Furnariidae) 29
Oviedo (Gonzalo Fernández de Oviedo y Valdés) 174–175
  La Natural Hystoria de las Indias (1526) 175
Owen, Denis 142
Owls (Strigidae and Tytonidae) 22
Oxfam 120
Oxford 120
Oxford University, UK 142

Pacific islands 11, 28, 39, 181, 196, 209, 223
Paella 241–243
Painted berrypeckers (Paramythiidae) 31, 35
Painted Honeyeater 102–103
Paleogene period 146
Palestine Sunbird 165, 172–173, 206, 209–210
Palms (Arecaceae) 173, 198
Panama 170
Pantanal, Brazil 78
Park, Geoff 230–231, Plate 25
  Natural Newstead blog 230–231

Parrots and their relatives (Psittaciformes) 4, 22, 28, 32, 36–37, 98, 103, 163, 193, 202, 219–220
Passenger Pigeon 172
Passerines or songbirds (Passeriformes) 14, 22, 28, 29, 33–36, 39, 42, 47, 50, 74, 74–75, 99–100, 129, 148–149, 157, 160, 164, 172–173, 192, 203, 220
Passionflowers 137–138, 139, 237
Paton, David 77
Pauw, Anton 25, 71–72, 137, 160, 203, 235
Pea and bean family or legumes (Fabaceae) 58, 64, 73–74, 131, 135, 158–159, 171, 198, 211
Peacocks 138
Pender, Richard 227, 236
Penduline-tits (Remizidae) 31
Penguins 7, 207
Penstemons (*Penstemon* and *Keckiella*) 52–54
Peru 17, 61, 82–85, 107, 153, 167–168
Pest control 186, 195–196
Peter, Craig 157
Phlox family (Polemoniaceae) 54, 63
Phylogenetics – see DNA, RNA and molecular systematics
Pigeons and doves (Columbidae, Columbiformes) 3, 4, 12, 20, 31, 33, 40, 171–172
Pikes Peak, Colorado, USA 105
Pincushion (*Leucospermum* sp.) 184, Plate 24
Pineapple family (Bromeliaceae) 47, 74
Plant behaviour 89–91, 92
Plant communities – see Habitats and plant communities
Plant Species Biology (journal) 85
Plantain family (Plantaginaceae) 53
Pliocene 11, 147
Poinsettia 198–199
Poland 11
Pollen
  Characteristics 67–69
  Feeding 55, 77–78, 143
  Grooming 68
  Heterospecific pollen transfer (HPD) 113–114
  Placement 4, 67–69, 84–85, 90
  Pollinia 69–72
  Robbing/collecting 69, 90, 98
  Viscin threads 68
Pollination as a process 3, 9, 82
Pollination syndromes 48–49, 50, 55, 58–63, 64, 92, 103, 131–133
  'Secondary pollinators' 62, 147
  Teleology 62
Pollination systems 17, 50–52, 63, 71, 74–75, 135, 160, 194, 210
  Bimodal 63
  Defined 49, 52, 59
  Generalised – see Generalist flowers and pollination systems
  Lability of 51–54
Pollinators & Pollination: Nature and Society (2021) 2, 17, 44, 59, 92, 146, 169, 204, 228, 231
Poppies 77
Porsch, Otto 94
Poulin, Brigitte 100
Praying mantis 4, 125
Prendergast, Kit 190, Plate 28
Price, Mary 16, 63, 133, 175
Prickly pear cactus 77, 242
Project Drawdown 240
Project Gutenberg 163
Prosser, Sean 101–102
Protea 122–123
Protea and banksia family (Proteaceae) 36, 47, 122–123, 198, 220, 239, Plates 4 & 28
Psychotria 205
Pterosaurs 8
Pueblo culture 169
Pumiliornis 12–13
Purple Sunbird 185–186
Purple-rumped Sunbird 185–186
Purple-throated Carib 111, 128
Pyke, Graham 76, 102–103

Quaternary 11
Queensland, Australia 11, 104, 190, 224, Plate 28

Rader, Romina 187
Raguso, Rob 98–99
Rainbow Lorikeet 36, 220, Plate 4
Ramíreza, Fernando 189
Rathcke, Beverly 93
Ratites (Paleognathae) 7, 21
Rats – see Mammals and flowers
Rech, André Rodrigo 107, 133, 244
Red Fody 239
Red List criteria – see International Union for Conservation of Nature (IUCN)
Red-legged Honeycreeper 189
Red-rumped Cacique 202
Red-whiskered Bulbul 239
Reed warblers and allies (Acrocephalidae) 30
Regan, Eugenie 19, 35
Regent Honeyeater 224
Reilly, John 20
  The Ascent of Birds (2019) 20
Remolina-Figueroa, Daniela 244
Ren, Zong-Xin 174, Plates 18, 19
Representations of flower-visiting birds
  Art works 167–170, 172–174, 181, Plates 17, 18, 19
  Coins and stamps 183–184
  Food and drink labels 184, Plates 22, 23, 24
  Guitars 182–183, 184, Plates 20, 21
  Music 182
Restoration – see Habitat and ecosystem restoration
Resoviaornis 10
Réunion 70–71, 149
Rewards for pollination 66, 72–79
Rhododendrons 25, 38, 153
Richea 32
Rico-Guevara, Alejandro 129, 138–139
Rio de Janeiro, Brazil 176
Rio Grande do Sul, Brazil 201
Ritchie, Alison 239
Robertson, Alastair 89–90
Robertson, Charles 214
Rocky Mountain Biological Laboratory (RMBL), USA 63, 133, 206
Rodents – see Mammals and flowers
Rodríguez-Gironés, Miguel 95
Roguz, Katarzyna 173
Rose family (Rosaceae) 14, 192
Rose, Jeffrey 54
Rosefinches (Fringillidae, subfamily Carduelinae) 34, 153
Rose-ringed (or Ring-necked) Parakeet 219
Roubik, David 194
  Pollination of Cultivated Plants in the Tropics (1995) 194
Royal National Park, Sydney 76
Royal Poinciana (see also *Delonix*) 198
Royal Society 61, 176
Rubega, Margaret 129
Ruby-crowned Tanager Plate 12
Ruby-throated Hummingbird 183, 214
Rufous Hummingbird 39, 101–102, 104–105, 206, Plates 22 & 23
Rufous-breasted Hermit 111, 128
Rufous-collared Sparrow 135
Rumphius, Georg Eberhard 179–180
  Herbarium Amboinense (1741–1750) 180

Sabino, William 193
Saccharomyces 118
Sachatamia Lodge, Ecuador 24
Saguaro 40, 49
  Cultural importance and fruit harvesting 193
Sahagún, Bernardino de 177
  Florentine Codex (*La Historia General de las Cosas de Nueva España* – 1545–1590) 177
Sahara 146, 240
Salvia 68, 205
Sandalwood order (Santalales), 89
Sangihe (or Sangir), Indonesia 223, 225

Sangihe White-eye 225
Santa Marta Mountains, Colombia 152
Santa Marta Sabrewing 152
Santamaría, Luis 95
Santore, Joey Plate 9
Sao Tome Sunbird 128
Sap feeding 36, 37, 40, 102
Sapsuckers (*Sphyrapicus*) 37
Sardinian Warbler 35, 159
Scandinavia 3, 167
Scarlet Gilia 63
Scarlet Honeyeater 230–231, Plate 25
Scarlet Myzomela – see Scarlet Honeyeater
Scarlet-chested Sunbird 1
Schlatmann, Alexander 41, Plates 1, 6, 27
Science (journal) 157
Scintillant Hummingbird Plate 26
Sclater's Myzomela 217
Scybalium 134, Plate 12
Seasonality 4, 55, 73, 75, 77, 89, 91, 100–101, 103–105, 112–114, 152, 202, 204
Seaweed 'pollination' 160
'Secondary pollinators' – see Pollination syndromes
Seed dispersal 40, 83, 91, 102, 172, 180, 189, 196, 209–210, 213, 222, 241
Seed predation (see also Granivory) 87, 135
Seedsnipes (Thinocoridae) 31, 40, 78, 135
Şekercioğlu, Çağan 195
Serra do Espinhaço, Brazil 113
Serra do Mar State Park, Brazil 204
Serrano-Serrano, Martha 55
Sérsic, Alicia 78–135
Sexual selection 121, 138–139
Seychelles 238–239
Seychelles Bulbul 239
Seychelles Sunbird 239
Sharpbill, Royal Flycatcher and allies (Oxyruncidae) 31
Shelley, George Ernest 182
A Monograph of the Nectariniidae, or Family of Sun-Birds (1876–1880) 182
Shenzhen, China 197, 200
Sherpa people 153
Shifting baselines in conservation 241–243
Shining Sunbird 172–173
Shiny Cowbird 202
Shipibo-Conibo people 169
Shorebirds – see Gulls and shorebirds (Charadriiformes)
Showy mistletoes (Loranthaceae) 89
Shrikes (Laniidae) 30
Sicklebill hummingbirds 126–128
Silberbauer-Gottsberger, Ilse 45–46, 55
Silva, Luís da 161, 164
Silva, Paulo 202
Silvereye or *Pihipihi* 64, 171, 190, 225
Singapore 200
Singh, Vineet Kumar 85
Skutch, Alexander 104
Sky islands 4, 151–154
Slipper flower (*Calceolaria* sp.) 78, 135
Songbirds – see Passerines or songbirds (Passeriformes)
Sonoran Desert, North America 192–193
Sophora 171
Souimanga Sunbird 212
South American Bushmaster 116, 125
South or southern Africa 17, 25, 48, 58–61, 65–66, 69–71, 74–76, 103, 107, 122–123, 136–137, 147, 183–184, 203, 205, 209–210, 235
Southern Double-collared Sunbird 209
Southern Water Vole 242
Spain 87, 119, 158–160, 209, 241–242, Plate 15
Spanish Sparrow 87
Sparrows – see Old World sparrows (Passeridae)
Spathodea 211–212
Specialisation 4, 14, 15, 16, 75, 83, 100, 102, 109–112, 126–140

Speckled Mousebird 80, 81
Spiderhunters – see Sunbirds and spiderhunters (Nectariinidae)
Spiders (Arachnida) 4, 40, 75, 100, 121, 123–125, 144, 164, 195
 'Bird-eating' 123–125
Spindalises (Spindalidae) 31
Spotted Snapweed 143
Squash and melon family (Cucurbitaceae) 86
St Lucia 127
Stamen bodies 77–78
Stanley, Dara 58
Starlings (Sturnidae) 29
State of the World's Birds report 221–222
Stebbins, Ledyard George 62
Stefan Vogel 146–149
Stelzer, Ralph 148
Stiles, Gary 127
Stitchbird (Notiomystidae) 31
Strelitzia reginae – see Bird of Paradise (plant)
Sugarbirds (Promeropidae) 31
Sugar-water feeders – see Bird feeders
Sulawesi 41, 223
Sun, Shi-Guo 192
Sunbirds and spiderhunters (Nectariinidae) 1, 2, 4, 12, 14, 17, 18, 27, 29, 33, 36, 39, 41, 47, 49, 65, 70–72, 75, 86, 88, 102, 104, 107, 108, 111–113, 123–125, 128–132, 136–137, 152–153, 157, 165, 172–173, 182–184, 185–186, 191–192, 203, 205, 206, 209–211, 212, 217, 219, 234, 239
 Absence on the Canary Islands 147–149
 'Cinnyridae' 182
 Conservation of 223–225, 234–235
 Hovering ability 25–26
 Taxonomy 23–24
Sunshine Wattle 64
Suriname 123, 183
SURPASS2 project 229–230
SVD or single-visit deposition experiments 82–83, 136
Sweden 159, 238
Swifts (Apodidae) and treeswifts (Hemiprocnidae) 14, 31, 39
Swinhoe's White-eye 104, 197, 200–201
Sword-billed Hummingbird 15, 102, 137–140, 152, 168
Sydney, Australia 11, 16, 35, 76, 220
Sylviid warblers (Sylviidae) 30, 35
Syria 165
Systematics – see Taxonomy and systematics
Sytsma, Kenneth 54

Tacazze Sunbird 1
Tamang people 153
Tanagers, honeycreepers, etc. (Thraupidae) 29, 34, 78, 79, 134, 135, 189, 193, 202
Tanzania 17, 86, 151–152, 224
Tasmania, Australia 32, 60
Taxonomy and systematics 20–24, 26, 59, 107, 159, 170, 182, 200, 245
Tayra 134
Tello-Ramos, Maria Cristina 94
Temeles, Ethan 127–128
Tenerife Blue Chaffinch 104, 239
Tenerife, Canary Islands 18, 65, 72, 119, 144–149, 208–209, 241–242
 Anaga Mountains 18, 145–146
 'Darwin's Unrequited Isle' 145
 Malpais de Güímar 241
Teratornis 10
Thailand 104
The Conversation 218
Thevet, André 176–177, 178
 Les Singularitez de la France Antarctique (1557) 176
 The New Found Worlde, or Antarctike (1568) 176
Thompson, John 128
Thornbills and allies (Acanthizidae) 30, 64
Thrushes and allies (Turdidae) 29
Thurston, Ginger 238
Tierra del Fuego, South America 47, 78, 165

Tit Berrypecker 35
Tits, chickadees and titmice (Paridae) 14, 17, 30, 42, 75, 134, 147, 160–162, 164, 196, 200
Tityras and allies (Tityridae) 30
Tohono O'odham people 192–193
Tongues adapted for nectar-feeding 35, 36, 37–38, 129, 135–136, 149, 174, 178, 209
Traveset, Anna 143–144
Tree Bumblebee 231–232
Tree Tobacco 25, 93, 208–211, 212, 215, 241–242
Tree-babblers, scimitar-babblers and allies (Timaliidae) 30
Trinidad and Tobago 183, 235
Trinity College Dublin, Ireland 41
Trogons (Trogonidae, Trogoniformes) 32, 44
Tropical Biology Association (TBA) 86, 130, 151
Troupials and allies (Icteridae) 29, 33
Trumpet Vine 183, Plate 20
Tucson, Arizona, USA 105
Tui or *Kōkō* 89, 171
Tumarae-Teka, Kirituia 171
Turquoise-throated Puffleg 223
Tyrant flycatchers (Tyrannidae) 27, 29
Tyrian Metaltail 83–84
UK Centre for Ecology and Hydrology 229

Uluguru Mountains, Tanzania 152
UNESCO 170, 183
UNESCO Kogelberg Biosphere Reserve, South Africa 183
United Kingdom 3, 6, 13, 50, 84, 148, 213, 229, 238
United Nations 194, 206, 234
United Nations' Decade on Ecosystem Restoration 234
Universidad Nacional de Salta, Argentina 50
Universitas Halu Oleo, Sulawesi 41
University of California, Riverside, USA 16
University of Campinas, Brazil 185, 198, 201
University of Connecticut, USA 122
University of Copenhagen, Denmark 115
University of Extremadura, Spain 158
University of New England, Armidale, Australia 218
University of Northampton, UK 57–58, 145
University of Sussex, UK 216
Urban and suburban birds and flowers 5, 36, 190, 197–207, 220
Urban bird diversity 199
Urbanisation as a conservation threat 202–203, 221, 230, 236
US National Science Foundation 133
USA 16, 103, 125, 169, 238, 242

Valido, Alfredo 147
van der Kooi, Casper 96
van der Niet, Timo 65
van Leeuwen, Willem Marius Docters 180
Vannette, Rachel 119–120
Variable Oriole 202
Variable Sunbird 1, 130
Venezuela 100
Verbena family (Verbenaceae) 133
Verdin 38
Vervet Monkey 80, 87
Victoria, Australia 230–231
Vietnam 192
Vigors, Nicholas Aylward 181–182
Vireos, shrikebabblers and *Erpornis* (Vireonidae) 30
Vonnegut, Kurt 200

Wagner, Helmuth 101
Wagtails and pipits (Motachillidae) 22, 30, 33
Wakatobi Islands, Sulawesi 41
Wallace, Alfred Russel 117
Walsh, Seana 236
Warbling White-eye 192, 200, 219
Waser, Nick 16, 38, 63, 95, 105, 107, 133, 156–157, 214,

Washington State University, USA 103
Wasps (Hymenoptera) 50, 125, 134, 135
Waterfowl (Anseriformes) 22
Watson, David 89
Wattlebirds (Callaeidae) 31
Watts, Stella 83–85, 107, 133, 153
Waxbills and allies (Estrildidae) 29
Waxwings 225
Weavers and allies (Ploceidae) 29
Wester, Petra 25, 68
Western Australia 47, 239
Western Honey Bee 171, 185, 224
Western Violet-backed Sunbird 23
Whelan, Christopher 195
White, Gilbert 97, 161–162, 163, 173
The Natural History and Antiquities of Selborne (1789) 161–162
White-chested White-eye 225
White-eyes, yuhinas and their relatives (Zosteropidae) 27, 29, 33, 70–71, 88, 107, 125, 165, 227
Conservation of 224–225, 239
Diet 75, 102
Invasive 219, 227
Taxonomy 29, 200
Urban 197, 200–201
Whitehead, Michael 118
Whiteheads (Mohouidae) 31
White-tipped Sicklebill 127
White-winged Dove 40, 49, 50
Wikipedia 19, 49, 90, 222, 241–242
Wild Chilli 196
Wilde, Volker 12
Wiley, Jim 149–51
Williams, Paul 153
Willows (*Salix* spp.) 14, 160–161, 166, 200
Wilson, Paul 52–53
Wind pollination 69, 187
Woodhoopoes and scimitarbills (Phoeniculidae, Bucerotiformes) 31, 40, Plate 7
Woodpeckers (Picidae) 4, 32, 37–38, 40, 41, 49, 88, 157, 178
Woodswallows, bellmagpies, and allies (Artamidae) 32
Woolly Mammoth 222
Wrens (Troglodytidae) 12, 22, 29
Wüest, Rafael 114

Xhosa people 235

Yangtze basin, Asia 104
Yeasts in nectar 4, 76, 100, 117–120
Yellow Fever Tree 80
Yellow Rattle 89
Yellow-bellied Sunbird-Asity 36
Yellow-breasted Chat (Icteriidae) 31, 34
Young, Allen 100–101
Young, Truman 41
YouTube 118, 183
Yucatan, Mexico 151
Yuccas 135
Yunnan Banana 191

Zanata, Thais 112